Easy Guide to Health and Safety

Endorsed by

RoSPA endorses this publication as offering a high quality introduction to the management of Health & Safety. It is particularly appropriate for those who own or are responsible for organisations with less than 250 employees. Although it can be used as a study aid for nationally recognised qualifications, RoSPA endorsement does not imply that this publication is essential to achieve any specific qualification. All responsibility for the content remains with the publisher.

Easy Guide to Health and Safety

Phil Hughes MBE
and
Liz Hughes

Endorsed by

AMSTERDAM • BOSTON • HEIDELBERG • LONDON • NEW YORK • OXFORD • PARIS
SAN DIEGO • SAN FRANCISCO • SYDNEY • TOKYO
Butterworth-Heinemann is an imprint of Elsevier

ELSEVIER

Butterworth-Heinemann is an imprint of Elsevier
Linacre House, Jordan Hill, Oxford OX2 8DP, UK
30 Corporate Drive, Suite 400, Burlington, MA 01803, USA

First edition 2008

Copyright © 2008, Phil Hughes and Liz Hughes. Published by Elsevier 2008.
All rights reserved

The right of Phil Hughes and Liz Hughes to be identified as the author of
this work has been asserted in accordance with the Copyright, Designs and
Patents Act 1988

Notice
No responsibility is assumed by the publisher for any injury and/or
damage to persons or property as a matter of products liability, negligence
or otherwise, or from any use or operation of any methods, products,
instructions or ideas contained in the material herein

Library of Congress Cataloging-in-Publication Data
A catalog record for this book is available from the Library of Congress

British Library Cataloguing in Publication Data
A catalogue record for this book is available from the British Library

ISBN: 978-0-7506-6954-2

For information on all Butterworth-Heinemann publications
visit our web site at books.elsevier.com

Printed and bound in Slovenia
08 09 10 10 9 8 7 6 5 4 3 2 1

Working together to grow
libraries in developing countries

www.elsevier.com | www.bookaid.org | www.sabre.org

ELSEVIER   BOOK AID
            International   Sabre Foundation

Contents

Preface

Are you worried about Health and Safety? Totally confused? Or just plain scared? Maybe you have given up on it altogether and are just trusting to luck. You most certainly are not alone. But stop worrying, help is at hand!

In 2003, Phil Hughes and Ed Ferrett, with a little help from Liz Hughes, wrote the original *Introduction to Health and Safety at Work* which quickly became a best-seller in its field. But after listening and chatting to people who are responsible for small businesses, Phil and Liz very soon became aware that there's also a huge need for something much simpler, something written, not for an exam syllabus (as *Introduction to Health and Safety at Work* was) but for people, probably like yourself, who are responsible for small and medium sized enterprises .

For the thousands of people who are running small businesses, health and safety seems like a very complicated issue that, if done properly, takes up a lot (too much in fact) of their valuable time. Are you one of those people? If you are, it's for you that this book has been written. We have divided it up into sections, so whatever business you are involved in, you should be able to find something that relates to your needs. We have provided clear information, ways of planning and ideas about where to go for help if you need more detail or specialisation.

So whether you are running a garage or a hair salon, a nursing home or a plant nursery, or any one of the thousands of small businesses that are part of the fabric of Britain today, you should be able to find help in this book. Maybe you are a builder who has suddenly discovered asbestos during a renovation? Or a beautician who finds that one of your staff is developing dermatitis? Or maybe a member of your admin staff is suffering from difficulties that seem to be connected with using a VDU? We can point you in the right direction, and help you to solve the problem.

Phil has provided the technical content of the book. He has had over 30 years world-wide experience as Head of Environment, Health and Safety in two large multinationals, Courtaulds and Fisons. He started in Health and Safety, in the Factory Inspectorate in the Derby District in 1969. He moved to Courtaulds Plc in 1974 and that year joined the Institution of Occupational Safety and Health (IOSH), becoming Chair of the Midland Branch, National Treasurer and in 1990–1991, President. He was very active with the National Examination Board in Occupational Safety and Health (NEBOSH) for over ten years and served as Chair from 1995 to 2001. He is a Professional Member of the American Society of Safety

Authors: Phil & Liz Hughes

Engineers and has lectured widely throughout the world. Phil received the Distinguished Service Award of the Royal Society for the Prevention of Accidents (RoSPA) in May 2001 and became a director and trustee of RoSPA in 2003. He received an MBE in the New Years Honours List 2005 for services to Health and Safety.

Liz gained an honours degree at the University of Warwick, later going on to complete a Masters degree at the same university. She taught psychology in further and higher education, where most of her students were either returning to education after a gap of many years, or were taking a course to augment their professional skills. She went on to qualify as a social worker specialising in mental health issues and was involved in the closure of long stay psychiatric hospitals in Warwickshire and the development of new housing and care units in the community. She later moved to the voluntary sector where she managed development for a number of years. Liz then helped to set up and manage training for the National Schizophrenia Fellowship (now called Rethink) in the Midlands. She wrote the Study Skills and Exam Technique for NEBOSH students and the Report Writing section for the *Introduction to Health and Safety at Work*.

But don't think for a moment that a life spent in Health and Safety needs to cramp your style! We are keen sailors – we lived on board our boat for 3 years and sailed in all kinds of wind and weather. We are also keen skiers, cyclists and walkers. Our four children are now adults and have children themselves and as well as instilling into our grandchildren the principles of safety, we are encouraging them to walk and cycle every-where, learn to sail and ski and generally enjoy a healthy, active lifestyle.

We would like to acknowledge the help given by Robert Kerkham who took a number of new photographs of unusual working situations; and thanks to Doris Funke of Elsevier and her team who supported the project throughout and have made this book possible. Definitions used by the relevant legislation and information from the Health and Safety Commission and Executive have been utilized. Most of the references used have come from the HSE books range of booklets which are considered to be the right level for those people managing a small business.

Liz and Phil Hughes
January 2008

Acknowledgements

Figure 1.1	Courtesy of RH Kerkham.
Figure 1.7	Courtesy of RH Kerkham.
Figure 1.10	Courtesy of RH Kerkham.
Figure 1.11	Courtesy of RH Kerkham.
Figure 1.12	Courtesy of RH Kerkham.
Appendix 1.3, example 1	Source HSE.
Figure 2.3	Courtesy of RH Kerkham.
Figure 3.3	Source HSE. © Crown copyright material is reproduced with the permission of the Controller of HMSO and Queen's Printer for Scotland.
Figure 3.4	Courtesy of RH Kerkham.
Figure 3.6	Redrawn from HSE flowchart at http://www.hse.gov.uk/smallbusinesses/flowchart.htm. © Crown copyright material is reproduced with the permission of the Controller of HMSO and Queen's Printer for Scotland.
Figure 4.1	Courtesy of Bridget and Mo Vyze, Dock o'the Bay Restaurant, 69, The Avenue, Southampton.
Figure 4.2	From HSG 76 *Health and Safety in Retail and Wholesale Warehouses* (HSE Books, 1992), ISBN 9780118857314. © Crown copyright material is reproduced with the permission of the Controller of HMSO and the Queen's Printer for Scotland.
Figure 4.3	Courtesy of RH Kerkham.
Table 4.1	From INDG 293 (rev1) *Welfare at Work* (HSE Books, 2007) ISBN 9780717662647. © Crown copyright material is reproduced with the permission of the Controller of HMSO and Queen's Printer for Scotland.
Table 4.2	From INDG 293 (rev1) *Welfare at Work* (HSE Books, 2007) ISBN 9780717662647. © Crown copyright material is reproduced with the permission of the Controller of HMSO and Queen's Printer for Scotland.

Figure 4.6   Redrawn from HSG 136 *Workplace Transport Safety* 2nd edition (HSE Books, 2005), ISBN 9780717661541. © Crown copyright material is reproduced with the permission of the Controller of HMSO and Queen's Printer for Scotland.

Figure 4.10   Redrawn from *A short guide to making your premises safe from fire* (DCLG, Publication code 05 FRSD 03546). © Crown copyright material is reproduced with the permission of the Controller of HMSO and Queen's Printer for Scotland.

Figure 4.11a   Redrawn from *A short guide to making your premises safe from fire* (DCLG, Publication code 05 FRSD 03546). © Crown copyright material is reproduced with the permission of the Controller of HMSO and Queen's Printer for Scotland.

Figure 4.11b   Redrawn from *A short guide to making your premises safe from fire* (DCLG, Publication code 05 FRSD 03546). © Crown copyright material is reproduced with the permission of the Controller of HMSO and Queen's Printer for Scotland.

Figure 4.12   Courtesy of Armagard.

Figure 4.15   Reproduced from INDG 236 *Maintaining portable electrical equipment in offices and other low risk environments* (HSE Books, 1996), ISBN 9780717612727. © Crown copyright material is reproduced with the permission of the Controller of HMSO and Queen's Printer for Scotland.

Figure 4.16   Courtesy of Stocksigns.

Figure 4.18   Reprinted from *Safety with Machinery* Second Edition, John Ridley and Dick Pearce, pages 26–34, 2005, with permission from Elsevier.

Figure 4.19   Courtesy of Draper.

Figure 4.22   Redrawn from L 23 *Manual Handling Operations – Guidance on Regulations* (HSE Books, 2004), ISBN 9780717628230. © Crown copyright material is reproduced with the permission of the Controller of HMSO and Queen's Printer for Scotland.

Figure 4.23   Redrawn from *Manual Handling in the Health Services* 2nd edition (HSE Books, 1998), ISBN 9780717612482. © Crown copyright material is reproduced with the permission of the Controller of HMSO and Queen's Printer for Scotland.

Table 4.3	Reproduced from INDG 225 (rev1) *Preventing slips and trips at work* Revised edition (HSE Books, 2005), ISBN 9780717627608. © Crown copyright material is reproduced with the permission of the Controller of HMSO and Queen's Printer for Scotland.
Figure 4.25	Redrawn from CIS 49 REV1 *General Access Scaffolds and Ladders. Construction Information Sheet No. 49* (revision) (HSE Books, 2003). © Crown copyright material is reproduced with the permission of the Controller of HMSO and Queen's Printer for Scotland.
Figure 4.27	Redrawn courtesy of Bratts Ladders – www.brattsladder.com.
Figure 5.6a	From INDG 223 (Rev3) *Managing Asbestos in Premises* (HSE Books, 2002), HSE ISBN 9780717625642. © Crown copyright material is reproduced with the permission of the Controller of HMSO and Queen's Printer for Scotland.
Figure 5.6b	From INDG 223 (Rev3) *Managing Asbestos in Premises* (HSE Books, 2002), HSE ISBN 9780717625642. © Crown copyright material is reproduced with the permission of the Controller of HMSO and Queen's Printer for Scotland.
Figure 5.8	From HSE website, hairdressing section. © Crown copyright material is reproduced with the permission of the Controller of HMSO and Queen's Printer for Scotland.
Figure 5.9	Redrawn from graph on HSE website, hairdresser section at http://www.hse.gov.uk/hairdressing/chart.htm. © Crown copyright material is reproduced with the permission of the Controller of HMSO and Queen's Printer for Scotland.
Figure 5.10	From HSE website at http://www.hse.gov.uk/hairdressing/workplace.pdf. © Crown copyright material is reproduced with the permission of the Controller of HMSO and Queen's Printer for Scotland.
Figure 5.11	From HSE website, drugs and alcohol section. © Crown copyright material is reproduced with the permission

	of the Controller of HMSO and Queen's Printer for Scotland.
Figure 5.12	Courtesy of RH Kerkham.
Figure 5.13a	Redrawn from *Everything you need to prepare for the new smokefree law on 1 July 2007* (HM Government, 2007) available at http://www.smokefreeengland.co.uk/files/everything_u_need_new_sf_law.pdf. © Crown copyright material is reproduced with the permission of the Controller of HMSO and Queen's Printer for Scotland.
Figure 5.13b	Redrawn from *Everything you need to prepare for the new smokefree law on 1 July 2007* (HM Government, 2007) available at http://www.smokefreeengland.co.uk/files/everything_u_need_new_sf_law.pdf. © Crown copyright material is reproduced with the permission of the Controller of HMSO and Queen's Printer for Scotland.
Appendix 5.2	Reproduced from http://www.smokefreeengland.co.uk/files/smokefree_policy.pdf. © Crown copyright material is reproduced with the permission of the Controller of HMSO and Queen's Printer for Scotland.
Appendix 5.3	Reproduced from http://www.smokefreeengland.co.uk/files/a5_sign_sf_premises.pdf. © Crown copyright material is reproduced with the permission of the Controller of HMSO and Queen's Printer for Scotland.
Figure 6.3	Redrawn from HSG 121 *A pain in your workplace? Ergonomic problems and solutions* (HSE Books, 1994) ISBN 9780717606689. © Crown copyright material is reproduced with the permission of the Controller of HMSO and Queen's Printer for Scotland.
Figure 6.4	From INDG 362 (rev1) *Noise at work – Guidance for employers on the Control of Noise at Work Regulations 2005* (HSE Books, 2005), ISBN 9780717661657. © Crown copyright material is reproduced with the permission of the Controller of HMSO and Queen's Printer for Scotland.
Figure 6.6	From INDG 362 (rev1) *Noise at work – Guidance for employers on the Control of Noise at Work Regulations 2005* (HSE Books, 2005), ISBN 9780717661657. © Crown copyright material is reproduced with the permission of the Controller of HMSO and Queen's Printer for Scotland.
Figure 6.7	Redrawn from INDG 175 (rev1) *Health Risks from Hand-Arm Vibration* (HSE Books, 1998) ISBN 9780717615537.

© Crown copyright material is reproduced with the permission of the Controller of HMSO and Queen's Printer for Scotland.

Figure 7.1 Courtesy of RH Kerkham.

Figure 7.2 Courtesy of Speedy.

Figure 7.3 Courtesy of RH Kerkham.

Figure 7.5b From HSG 185 *Health and Safety in Excavations* (HSE Books, 1999), ISBN 9780717615636. © Crown copyright material is reproduced with the permission of the Controller of HMSO and Queen's Printer for Scotland.

Figure 7.6 Courtesy of Stocksigns.

Figure 7.8 From HSG 149 *Backs for the future: Safe manual handling in construction* (HSE Books, 2000), ISBN 9780717611225. © Crown copyright material is reproduced with the permission of the Controller of HMSO and Queen's Printer for Scotland.

Figure 7.9 From HSG 149 *Backs for the future: Safe manual handling in construction* (HSE Books, 2000), ISBN 9780717611225. © Crown copyright material is reproduced with the permission of the Controller of HMSO and Queen's Printer for Scotland.

Figure 7.10 Redrawn from HSG 150 (rev2) *Health and Safety in Construction* (HSE Books, 2006), ISBN 9780717661824. © Crown copyright material is reproduced with the permission of the Controller of HMSO and Queen's Printer for Scotland.

Figure 8.2 Courtesy of RH Kerkham.

Figure 8.6 From BI 510 *Accident Book* (HSE Books, 2003), ISBN 9780717626038. © Crown copyright material is reproduced with the permission of the Controller of HMSO and Queen's Printer for Scotland.

Table 8.1 Reproduced from INDG 214 *First aid at work: Your questions answered* (HSE Books, 1997), ISBN 9780717610747. © Crown copyright material is reproduced with the permission of the Controller of HMSO and Queen's Printer for Scotland.

Appendix 8.3 Courtesy of Stocksigns.

What is health and safety all about?

1

Summary

- Why health and safety is important to you and your business

- What hazard, risk and control is all about

- What a risk assessment is and how to conduct a simple example; plus

- A prompt list of hazards

- Completed examples of risk assessments

1.1 Introduction

In the UK accidents kill over 8000 people a year. Over six million people are injured, many of them seriously. More than 2 million people suffer from work related ill health. Yet we could often very easily prevent the actions and events that cause accidents and ill health, whether we are at home, at leisure, travelling or at work. This book looks at the workplace and explains how you can manage and control things so that people are protected and you, as a manager, employer, self-employed or employee can comply with the law. It is about helping you to live and work free of injury and ill health caused by your occupation.

Everyone needs to work safely and this book is intended for managers as well as the people working for them. It is designed to be an easy guide to health and safety covering in simple straightforward language the issues involved in many small businesses, and that means most of the enterprises in Britain today.

Falling down stairs, slipping on wet floors, hitting or being hit by objects, breathing in dangerous fumes, receiving an electric shock and getting burnt or killed in a fire are all common examples of the sorts of accidents that happen to people at work (Fig. 1.1).

Figure 1.1 Glass blower without proper protection at risk of serious burns.

This book shows the kinds of things which cause the more common accidents and health problems. It lets people see what applies to their work activities, and tells them how to get more help and information.

This is especially important for people in charge of work activities, for example people who employ others, because they have significant legal responsibilities. Everyone at work has legal responsibilities but they fall much more heavily on those in charge.

Throughout the book useful and well-produced Health and Safety Executive (HSE) booklets are referred to. You can often get these free from HSE Books or download them from the HSE website (www.hse.gov.uk). The HSE is the Central Government funded body that looks after health and safety at work.

For more sources of information see Chapter 9.

1.2 Why is health and safety such an important topic?

Nobody chooses to get hurt at work, yet every year over 300 people are killed at work. About 160,000 non-fatal accidents are reported and some 2.2 million people are estimated to suffer from ill health caused or made worse by work. If those who are killed and injured on the roads, in mobile workplaces, are added, this amounts to over 1,000 deaths and 260,000 injuries. That is over 100 people killed and 36,000 injured every month. Putting other peoples' lives and health at risk in this way is not acceptable, especially when, as is often the case, workers are not entirely aware of the risks they are taking. Every one should be able to go to work, feeling confident that they will end each working day without being harmed.

It's easy to believe that accidents at work are unusual or exceptional, the sort of things that never happen in your workplace. But this is not true. What is true, is that a few basic measures could, in most cases, have prevented an accident from happening (Fig. 1.2).

Making your workplace safe doesn't have to be expensive, time consuming or complicated. In fact, working safely and efficiently will often save you money. But even more importantly, it can save lives.

The HSE has set out the main principles for sensible health and safety at work. Their press release said the following.

'Get a life', says HSC

Figure 1.2 Dangerous electrics on cement mixer – the proper plug had been lost.

The Health and Safety Commission (HSC, *a small central body with representatives from employers, unions and others like local authorities who have overall responsibility for the HSE and approving new legislation*) has urged people to focus on real risks – those that cause real harm and suffering – and stop concentrating effort on trivial risks and petty health and safety. To help take this forward the HSE has launched a set of key principles: practical actions that we believe sensible risk management should, and should not, be about.

Launching the principles at a children's sailing centre in north London, Bill Callaghan, former Chair of the HSC, said: 'I'm sick and tired of hearing that 'health and safety' is stopping people doing worthwhile and enjoyable things when at the same time others are suffering real harm and even death as a result of mismanagement at work'.

'Some of the "health and safety" stories are just myths. There are also some instances where health and safety is used as an excuse to justify unpopular decisions such as closing facilities. But behind many of the stories, there is at least a grain of truth – someone really has made a stupid decision. We're determined to tackle all these. My message is that if you're using health and safety to stop everyday activities – get a life and let others get on with theirs.'

Lending support to the principles, author and adventurer Ben Fogle said: 'Children encounter risk everyday and its important that, through activities like those being carried out today, they learn how to enjoy themselves but also stay safe.

'I probably take more risks than most – and I wouldn't want my life to be any other way. No one wants a world where children, in fact anyone, is wrapped in cotton wool, prevented from taking any risks and scared of endeavour.' 'That' why I'm supporting HSE's launch and am happy to endorse these principles'.

Sensible risk management **IS** about:

- ◆ ensuring that workers and the public are properly protected;
- ◆ providing overall benefit to society by balancing benefits and risks, with a focus on reducing real risks – both those which arise more often and those with serious consequences;
- ◆ enabling innovation and learning, not stifling them;
- ◆ ensuring that those who create risks manage them responsibly and understand that failure to manage real risks responsibly is likely to lead to robust action; and
- ◆ enabling individuals to understand that as well as the right to protection, they also have to exercise responsibility.

Sensible risk management **IS NOT** about:

- ◆ creating a totally risk free society;
- ◆ generating useless paperwork mountains;
- ◆ scaring people by exaggerating or publicising trivial risks;
- ◆ stopping important recreational and learning activities for individuals where the risks are managed (Fig. 1.3a and 1.3b); and
- ◆ reducing protection of people from risks that cause real harm and suffering.

Figure 1.3a Yachts – well managed risks.

Figure 1.3b Car rally – well managed risks.

Commenting on the principles Jonathan Rees, HSE Deputy Chief Executive, said: 'We want to cut red tape and make a real difference to people's lives. We are already taking action to put the principles into practice. Last month we published straight-talking guidance on risk management, but we cannot do this alone. That's why I welcome the broad alliance of support for this initiative – organisations representing employers, workers, insurers, lawyers, volunteers, health and safety professionals and many others who have made positive contributions to our approach'.

'These principles build on all of this and will hopefully drum home the message that health and safety is not about long forms, back-covering, or stifling initiative. It's about recognising real risks, tackling them in a balanced way and watching out for each other. It's about keeping people safe – not stopping their lives'.

It should of course be remembered that many things change with time. What was acceptable 30 years ago is not acceptable today. In the early 1970s we had four young children who were packed into a medium-sized standard car. Despite working in the central safety department of a very large company I could not persuade them to provide, even an estate car. Even in the 1980s our children steadfastly refused to wear rear seat belts through town 'in case their friends saw them'. On one family outing I had to refuse to go further unless they fastened their seat belts.

Those same children now have their own families and refuse to move a 100 yards without a child seat in the car, even when far from home, in

Turkey, on holiday. Helmets on bikes is now the norm, our grandson's face was saved by one in France in 2005 (Fig. 1.4).

Figure 1.4 Jack and Grandpa.

We have watched our eldest grandson playing rugby in local competitions at 6, unheard of several years ago. Tag rugby has been introduced where players have a tape each side attached to a velcro belt. Grabbing the tape and pulling it off, is a tackle. They do not have scrums or intentional body contact. Each team has its own coach on the field to marshal the players. Those children had a tough competition, played hard, got bumped a bit but not hurt. No one puts tag rugby down to silly safety standards; quite the contrary. It is safety control standards allowing younger children to play rugby. At the same time top rugby players are seriously discussing the ever increasing injury toll to senior players. Each top team on average has one player a season whose rugby career is finished by injury. Here the sport's leaders are considering what should be done to combat the rise in serious injuries.

Health and safety at work is similar in that standards change with time and place. For example in 2008 it is common place for people to consider noise and vibration issues at work in European countries. In the late 1960s and 1970s people were unconcerned about these hazards. However in some developing countries, where poverty is severe, getting food to feed your family and preventing child prostitution is much more important, even in the 21st century, than preventing future loss of hearing.

1.3 What is health and safety all about?

Health and safety is about preventing people from being harmed at work, by taking the right precautions and by providing a satisfactory working environment.

These things are morally and ethically important, but there is more to it than that. A healthy and safe working environment can mean increased productivity, reduced staff turnover and lower staff sickness absence. Healthy staff, who feel that their safety is well looked after, are likely to be far happier and have a positive attitude to their work.

The costs of injuries, damage to plant, down time, insurance and missed dead lines can impose a high penalty on a business. A single incident can ruin a small company and cause huge disruption in a medium-sized business. And for self-employed people, the absence of just one person can cause the business to shut down (Fig. 1.5).

Figure 1.5 Chimney sweeps – one accident would close the business.

There are also important legal duties placed on employers, the self-employed and employees. These will be covered in later chapters. Breaches of these duties can result in fines, enforcement notices and even, in rare cases, imprisonment. These duties are enforced by the Health and Safety Executive (HSE) or the local authorities' Environmental Health Officers. These organisations are also there to help and can be contacted for both verbal and written advice.

☞ *'An Introduction to health and safety'* INDG259Rev1
☞ *'The HSE and You'* HSE37
☞ *'Need help on health and safety?'* INDG322

1.4 Role and function of other external agencies

The Health and Safety Commission, Health and Safety Executive and the Local Authorities (a term used to cover County, District and Unitary Councils) are all agencies external to the workplace or organisation, which have a direct role in the monitoring and enforcement of health and safety standards. But there are three other external agencies which have a significant influence on health and safety standards at work (Fig. 1.6).

Fire and rescue authorities

Environment agency

Insurance companies

HSC/HSE/Local authorities

Figure 1.6 Influences on the workplace.

1.4.1 Fire and rescue authority

Fire and rescue authorities are part of their local authority. They are normally associated with fire fighting, rescuing people in road and other accidents and giving general advice. They have also been given powers to regulate fire precautions within places of work under fire precautions law.

1.4.2 The Environment Agency (Scottish Environmental Protection Agency)

The Environment Agency was established in 1995 and was given the duty to protect and improve the environment. It is the central regulatory body for environmental matters and also has an influence on health and safety issues.

1.4.3 Insurance companies

Insurance companies play an important role in the improvement of health and safety standards. Since 1969, employers must, by law, insure against liability for injury or disease to their employees arising out of their employment. This is called employers' liability insurance. Some public sector organisations are exempted from this because any compensation is paid from public funds. Other forms of insurance include fire insurance and public liability insurance (to protect members of the public).

Premiums for all these types of insurance are related to levels of risk. This, in turn, is often related to standards of health and safety. In recent years, there has been a considerable increase in the number and size of compensation claims which has put insurance companies under pressure. So now, insurance companies are having a powerful influence on health and safety standards because they will work out your premium based partly on your fire and safety record.

1.5 Getting started – hazards, risk assessment and control

First of all you will need to understand how to identify hazards and minimise risks. This involves carrying out a risk assessment and then putting the right precautions into place. This may sound daunting but in practice in most businesses it is a fairly simple task.

A hazard is anything that can cause harm. For example things like electricity, chemicals, working from ladders and so on are hazards (Fig. 1.7).

Figure 1.7 Painter at work from a ladder.

A risk is the chance, high, medium or low that somebody will be harmed by the hazard. For instance, properly maintained computer equipment presents a low risk of electric shock, but if you are using naked flames where there are highly flammable liquids, like petrol, there will be a high risk of injury from an explosion or flash fire.

The expression 'risk control measures' means the same as 'health and safety precautions' Both are used to reduce the risk of injury or ill health.

Questions you need to ask to assess the risk:

◆ What is the activity? Where is it?
◆ How many people are exposed to the hazard?
◆ How often are they exposed to it?

- ◆ For how long are they exposed to it?
- ◆ How severe is it? What harm could occur?

So the assessment of risk involves looking at the activity, the workplace, the likelihood that harm will occur and the severity or the consequence of the harm.

Health and safety law says you must make a general risk assessment of your workplace. There are also more specific risk assessments required by different legislation; for example when lifting heavy objects, and when using dangerous chemicals. In many workplaces and situations e.g. offices and small shops the general risk assessment may be enough.

1.6 How do you make a risk assessment?

In most firms in the service or light industrial sectors or in office accommodation hazards are limited and simple. Identifying the hazards is not complicated and the risk assessment is nothing more than simply thinking about what could harm people and whether the precautions are adequate.

Every day when we overtake on the road we have to make risk judgements. We must consider oncoming traffic, the size of the vehicle in front, the space available to overtake, any vehicles behind, the condition of the road, weather conditions, visibility, pedestrians nearby, road restrictions and markings. And really, this is all that you will need to do in your place of work – consider what could cause harm, how likely it is to happen, how severe the consequences would be and how to reduce or control them.

The HSE has suggested a simple five step approach to risk assessment which involves:

1.6.1 Step 1: identifying hazards

Walk around your workplace and look at what could reasonably be expected to cause a health and safety problem. Ignore the trivia and concentrate on real hazards which could result in serious harm or affect several people. Ask peoples' opinions – they may have noticed things that are not immediately obvious. Manufacturers' instructions or data sheets can also help you to spot hazards and put risks into perspective. Also look at accident, and ill-health records.

There is a list of hazard prompts at the end of the chapter. Here are a few broad categories of hazard for you to look for:

(a) *Mechanical hazards*:

◆ moving machinery, for example a circular saw, where people could be trapped in drive belts, cut in contact with the rotating blades or struck by a piece of wood that is ejected out of the saw;

◆ mobile equipment, for example a tractor, where people could be crushed when run over, entangled in moving parts of the engine or drive shaft or the driver could be crushed if the vehicle turned over (Fig. 1.8);

◆ a paper guillotine where fingers could be cut in the shearing action of the knife; and

◆ a sewing machine where fingers can be injured in the stabbing and puncture movement of the needle.

Figure 1.8 Mechanical hazards mobile plant.

(b) *Physical hazards*:

◆ slipping on a wet floor or tripping on uneven flooring or trailing cables; and

◆ burns from a fire or cooking equipment or scalding from a hot cup of coffee (Fig. 1.9).

Figure 1.9 Cooker.

(c) *Chemical hazards*:

 ◆ spillage of corrosive acid which can severely damage skin and eyes through contact;
 ◆ toxic chemicals which can damage through being swallowed (ingestion) or breathed in (inhalation) or solvents that can also be absorbed through the skin (absorption) (Fig. 1.10); and
 ◆ very fine dusts that can be breathed in, for example asbestos.

Figure 1.10 Yesteryear chemist shop – with numerous chemicals.

Chemicals either in gas, vapour or liquid forms, have different effects on people. Some chemicals work very quickly and may cause asphyxiation. For example, carbon dioxide asphyxiates (chokes) people because they are starved of oxygen whereas cyanide poisons people very fast, if swallowed. Solvent vapours when breathed in overcome people so that they initially appear drunk and may die if exposure is prolonged and severe.

Other substances, such as fine asbestos dust have more of a mechanical effect on the body. It can damage the lungs by causing scar tissue to form inside them (fibrosis of the lung known as Asbestosis) and/or cause cancer. Both usually, but not always occur after many years of exposure to asbestos dust.

(d) *Environmental hazards*:
   ◆ in the global environment water can be polluted by chemicals and/ or by toxic waste. Something as simple as pouring old engine oil down the drain will have this effect;
   ◆ air can be polluted from burning rubbish or emissions from chimneys;
   ◆ the environment can be polluted by waste and rubbish left lying about to rot or contaminate;
   ◆ inside buildings there may be poor ventilation so that people become drowsy and unable to operate machines safely. In extreme conditions they may be overcome by lack of oxygen or build up, for example of carbon dioxide;
   ◆ excessively hot working conditions can cause heat exhaustion and hyperthermia;
   ◆ cold conditions may cause hypothermia and in extreme cold there is a danger of frost bite; and
   ◆ poor lighting with reflection and glare or simply too little light can also be classed as environmental hazards (Fig. 1.11).

(e) *Biological hazards*:
   ◆ this includes bacteria and micro organisms of the type encountered when working with animals or infected people; handling waste materials (particularly in hospitals) or working in contaminated environments (Fig. 1.12);

(f) *Organisational hazards include*:
   ◆ working with unreasonable timescales or excessive workloads;
   ◆ using an unsafe way of working where instructions are not provided or not followed;

Figure 1.11 Boat building Poor conditions.

Figure 1.12 Working with animals.

◆ poor supervision leading to untrained or inexperienced people making mistakes;

◆ working unsocial hours (as in shift work) resulting in high levels of stress;

◆ local difficulties between individuals because of a clash of culture, religion , race or personality; or

◆ the possibility that the internal structure of the organisation makes it difficult for people to feel in control of their working lives.

It is important to remember that some hazards are not obvious and it may not be possible to see, feel, hear or smell them until it is too late.

Examples of this are:

- colourless and odourless gases such as carbon dioxide and carbon monoxide;
- ultra violet (UV) and infra red (IR) and microwave radiation; and
- electricity.

Hazards from poor posture, due to bad seating, stress from work overload or poor working conditions may not be quickly or easily identified.

Other hazards may have to be learned when people start work, so you will need to explain the dangers of chemicals, toxic or unpleasant dusts, noise or vibration, where they apply to your business.

1.6.2 Step 2: Deciding who might be harmed and how this could happen

This includes employees, cleaners, visitors, contractors, maintenance workers, employees away travelling or at another workplace.

Members of the public and other people who may share a workplace should also be considered.

Young workers, trainees, new and expectant mothers may be particularly at risk.

1.6.3 Step 3: Evaluating the risks and deciding whether the existing controls are sufficient

Draw up an action list of work that has to be done giving priority to high risks and/or risks which could affect most people. Consider:

- Can the hazard be **removed altogether**?
- If not, how can the risks be **controlled** so that harm is unlikely?
- How **severe** is the harm likely to be?
- How **likely** is it to happen?

These are the questions you will need to ask yourself. Then you will be able to decide how much you will need to do to reduce the risk. Even when you have taken all the precautions some risk usually remains. At that point you will have to decide whether the remaining risk is high, medium or low. This is sometimes called the residual risk.

You will, of course, have to make sure you are complying with the law, but we shall tackle that later in the book (see Chapter 3).

Now ask yourself whether the standards and level of risk in your workplace are generally acceptable in your profession or industry.

There is a set of principles to help you decide whether the existing controls are sufficient. It is called the 'hierarchy of risk control measures'.

1.6.4 Hierarchy of risk control measures

If you really feel that you cannot completely eliminate a hazard, the next step is to control the risks so that any harm to the people involved with it is less likely. The following list has been put together to help you to achieve this. The points have been made in order of preference:

- ◆ use a less risky system, process or piece of equipment;
- ◆ prevent people gaining access to the hazard by barriers or enclosures;
- ◆ limit the amount of time that people are exposed to the hazard;
- ◆ organise the work so that people are the least exposed to the hazard and control the way in which people may be exposed to a risk. This involves safe systems of work and in some cases written permits to work in special high-risk situations such as entry into confined spaces;
- ◆ provide personal protective clothing; and
- ◆ make available adequate and suitable welfare facilities such as washing facilities and first aid.

1.6.5 Risk matrix chart

A risk matrix chart can be used to help to establish whether the residual risk is High, Medium or Low. Where high-risk activities have been identified, action should be taken immediately

$$\text{Risk} = \text{Severity} \times \text{Likelihood}$$

1.6.6 Likelihood of an incident occurring

The categories used for the **Likelihood** of an incident occurring are:

1. **Improbable.**

2. **Unlikely.**

3. **Even Chance.**

4. **Likely.**

5. **Almost Certain.**

1.6.7 Severity/consequences of an incident

Following the process of identifying the likelihood of an incident the severity or the consequence needs to be established using the same matrix chart. The categories used are:

1. Negligible – Non or trivial injury/illness/loss
2. Slight – Minor injury/illness – immediate first aid only/slight loss.
3. Moderate – Injury/illness 3 days or more absence (reportable* category) moderate loss.
4. High – Major injury (reportable* category)/severe incapacity/serious loss
5. Very High – Death(s)/permanent incapacity/widespread loss.

*Reportable means that they are incidents which are reportable to the HSE or local authority. See later chapters for further information.

1.6.8 Risk matrix chart

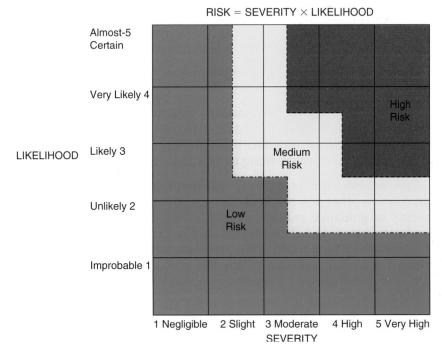

Figure 1.13 Risk martix chart.

1.6.9 Step 4: Keeping records

If you have 5 or more employees you must record the significant findings of the risk assessment. This means writing down the significant hazards and conclusions. The proforma can be used to record these findings.

Employees must be told about the findings.

Risk assessments have to be **suitable and sufficient.** You will need to show that a proper check was made; you asked about who was involved; you dealt with all the obvious hazards; the precautions are reasonable; and the residual risk low. To make things simple refer to procedures, documents, policies, etc.

You do not need to record insignificant risks. If your company has guides or manuals where hazards are listed and controls set out these can be referred to rather than spelling them all out again.

1.6.10 Step 5: Keeping you risk assessments up to date

New equipment or chemicals or procedures are bound to be brought into the workplace eventually. The risk assessments need to be revised when this happens. Trivial changes need not affect the risk assessment. If something new introduces significant risks you will have to review your assessment.

A routine check to ensure that precautions are working fine is sensible in any case.

You will often see the term **'Reasonably practicable'** used in health and safety, It means that there should be a balance between the effort, cost, inconvenience or time involved in setting up the workplace precautions and how much benefit they bring in reducing risks to people at work.

The risk assessment is best carried out by people who work in the area, as they know what the problems are and what dangers they face. There is plenty of guidance available from the HSE on particular types of risk or industry sectors to help with the task. In some cases you may feel you need inside help from health and safety advisers or safety representatives. If there are special problems or the key people have insufficient knowledge or experience, you may feel you need to bring in some outside expertise.

Risk assessments must be suitable and sufficient, **not perfect.**

☞ *Five steps to risk assessment* INDG163(rev1) HSE Books

Appendix 1.1

Hazard prompts

These hazard prompts can be used to help you make decisions. They are fairly extensive but not exhaustive.

1. Hazards associated with plant and equipment (including non-powered plant and hand tools).

Category	Type of harm	Examples of hazard
Mechanical	Trapping (crushing, drawing in and shearing injuries)	Two moving parts or one moving part and a fixed surface Conveyor belt and drive Vee belt and pulley Power press 'In-running nips' Mangle Guillotine Scissors Stapler Using hammer
	Impact (includes puncture)	Something that may strike or stab someone or can be struck against Moving vehicle Robot arm Sewing machine Drill Hypodermic needle Pendulum Crane hook
	Contact (Cutting, friction or abrasion)	Something sharp or with a rough surface Knife, chisel, saw, etc. Blender blade Circular saw blade Sanding belt Abrasive wheel Hover mower blade
	Entanglement (rotating parts)	Drill chuck and bit Power take off shaft PO threading machine Abrasive wheel
	Ejection (of work piece or part of tool)	Cartridge toot Thicknessing machine Using hammer and chisel Abrasive wheel

Category	Type of harm	Examples of hazard
Electrical	Shock/burn/fire/explosion	Electricity above 240V
		Electricity – 240Y
		Electricity – 110YCTE
		Extra low volt electricity
	Ignition sources	Static
		Batteries
Pressure	Release of energy (Explosion/ injection/implosion)	Compressed air
		Compressed gas
		Steam boiler
		Vacuum
		Hydraulic system
Stored energy	Flying/Falling materials	Springs under tension
		Springs under compression
		Hoist piatform/lift cage
		Conveyor tension weight
		Raised tipper lorry body
		Counterweight
		Load carried by crane
Thermal	Burns/fires/scalds/frostbite	Hot surface
		Using blow lamp
		Welding flame/arc
		Refrigerant
		Steam
Ionising radiation	Burns, cancer	X-rays thickness gauges using Gamma or Beta rays
Non ionising radiation	Burns	Micro wave
		Radio frequency
		Laser beams
		Ultra violet
		Infra red
Noise	Hearing loss, tinnitus, etc.	Noise $> 85\,dB(A)\ L_{EPd}$
Vibration	Vibration white finger, whole body effects.	Pneumatic drill
		Operation of plant
Stability	Crushing	Inadequate crane base
		Fork lift truck on slope
		Machine not bolted down
		Mobile Scaffold too high
		Scaffold not tied
Overload or defective due to mechanical failure	Crushing	Crane overload
		Chain sling
		Eye bolt overload
		Scaffold overload
		Hopper overfill

2. Hazards associated with materials and substances

Category	Type of harm	Examples of hazard
FIRE/EXPLOSION Combustion	Burns	Timber stack Coal store Paper store Grease Magnesium Straw Plastic foam
Increased Combustion	Burns	Oxygen enrichment
Flammable substance (inc. Highly and Extremely Flammable) See also explosive below	Burns	Petrol Propane gas Methane Carbon monoxide Methanol Paraffin Acetone Toluene
Oxidising substance	Burns	Organic peroxide Potassium permanganate Nitric acid Explosive material Fireworks Proprietary explosives Detonators Some oxidising agents Highly flammable gas in confined space
Dust explosions	Burns	Coal dust Wood dust Aluminium powder Flour
HEALTH HAZARDS Corrosive/irritating materials	Skin effects	Sulphuric acid Caustic soda Man made mineral fibre
Particles	Lung effects	Asbestos fibres Silica dust Dust mite faeces Pigeon droppings Coal dust Grain dust Wood dust

Category	Type of harm	Examples of hazard
Fumes	Acute and chronic effects on health (Local and systemic effects)	Lead fume Rubber fume Asphalt fumes
Vapours	Acute and chronic effects on health	Acetone 1,1,1-Trichloroethane Dichloromethane Benzene Isocyanates
Gases	Acute and chronic effects on health	Carbon monoxide Hydrogen sulphide Carbon disulphide Sulphur dioxide
Mists	Acute and chronic effects on health	Oil mist Printing ink mist Water – legionella
Asphyxiants	Acute and chronic effects on health	Nitrogen Carbon dioxide Argon
Health hazards by ingestion	Burns to upper alimentary tract	Toxic, harmful, corrosive and irritant liquids
	Poisoning	All harmful aerosols Polluted water Contaminated food and drink
Hazards by contact	Cuts, abrasions	Swarf Rough timber Concrete blocks
	Burns, Frostbite	Molten metal Frozen food

3. Hazards associated with the place of work

Category	Type of harm	Example of hazard
Pedestrian access	Tripping, slipping	Damaged floors Trailing cables Oil spills Water on floors Debris Wet grass Sloping surface Uneven steps Changes in floor level
Work at heights	Falls	Fragile roof Edge of roof Edge of mezzanine floor Work on ladder Erecting scaffold Hole in floor
Obstructions	Striking against	Low headroom Sharp projections
Stacking/storing	Failing materials	High stacks Insecure stacks Inadequate racking Stacking at heights
Work over/near liquids, dusts, grain, etc.	Fall into substances, drowning, poisoning, suffocation, etc.	Grain silo Tank Reservoir Sump Work over river Work near canal
Emergencies	Trapping in fire	Locked exits Obstructed egresses Long exit route

4. Hazards associated with the working environment

Category	Type of harm	Example of hazard
Light (*Note*: Also increases risk of contact with other hazards)	Eye strain, arc eye and cataracts	Glare Poor lighting Stroboscopic effect Arc welding Molten metal
Temperature	Heat stress, hypothermia	Work in furnace Cold room
	Heat stress, sunburn, melanoma, hypothermia, etc.	Outdoor work Hot weather Cold weather Wind chill factor Work in rain, snow, etc.
Confined spaces	Asphyxiation, explosion, poisoning, etc.	Work in tank Chimney Pit Basement Unventilated room Vessel Silo
Ventilation	'Sick Building Syndrome', nausea, tiredness, etc.	Fumes Odours Tobacco smoke

5. Hazards associated with methods of work

Category	Type of harm	Example of hazard
Manual handling	Back injury, hernia, etc.	Lifting Lowering Carrying Pushing Pulling Hot/cold loads Roughloads Live loads – animal/person
Repetitive movements	Work related upper limb disorders	Keyboard work Using screwdriver Using hammer and chisel Bricklaying Plucking chickens Production line tasks
Posture	Work related upper limb disorders, stress, etc.	Seated work Work above head height Work at floor level

6. Hazards associated with work organisation

Category	Type of harm	Example of hazard
Contractors	Injuries and ill health to employees by contractors work	Work above employees Use of harmful substances Contractors' welding
	Injuries and ill health to contractors' employees by work in premises	Process fumes Services (e.g. underground electricity cables) Stored hazardous materials
Organisation of work	Injuries to employees	Monotonous work Stress Too much work Lack of control of job Work too demanding
Work in public areas	Injuries and ill health of public	Trailing cables Traffic/plant movement Obstruction to blind person Obstruction to prams, etc Work above public

7. Other types of hazard

Category	Type of harm	Example of hazard
Attack by animals	Bite, sting, crushing, etc.	Bees Dog Bull Fleas Snake
Attack by people	Injury, illness, post trauma stress disorder	Criminal attack Angry customer Angry student Drunken person Drug abuser Mentally ill person
Natural hazards	Various injuries, illnesses	Lightning Flash flood

Appendix 1.2

Risk assessment record sheet

RISK ASSESSMENT RECORD

OPERATIONAL AREA:

DATE: UPDATED: UPDATED:

ASSESSOR:

Activity location	No. of people	Frequency of activity	Hazards	Existing controls	Initial risk factor	Risk elimination reduction	Residual risk factor

Activity location	No. of people	Frequency of activity	Hazards	Existing controls	Initial risk factor	Risk elimination reduction	Residual risk factor

Appendix 1.3

Example of risk assessments completed using a slight variations on the record forms

We have produced some example risk assessments to help you see what a risk assessment might look like. Example 1 comes from the HSE. We hope it is clear that a risk assessment should be about identifying practical actions that protect people from harm and injury, not a bureaucratic experience. We believe that for the great majority of risk assessments, short bullet points work well.

If you are involved in the same activities as those covered in the examples, you will find much of the detail directly relevant to you. However, you should not simply read across to your own business – these examples do not provide you with a short cut to your own assessment. All businesses have their unique features and a particular example may cover some hazards you do not have to deal with in your own workplace, and not mention some you do – you will have to take your own 5 steps when carrying out your own risk assessment. Even where the hazards are the same, the control measures you adopt may have to be different from those in examples so as to meet the particular conditions in your workplace.

Though each example deals with very different activities, you will see that there is a common approach.

Example 1

How was the risk assessment done?

The manager followed the guidance in '5 Steps to Risk assessment'.

1. To identify the hazards, the manager:
   • Read HSE's Office health and safety web pages, the Officewise leaflet and the more general Essentials of health and safety at work publication, to learn where hazards can occur.
   • Walked around the office noting things they thought might pose a risk, taking into consideration what they learnt from HSE's guidance.
   • Talked to supervisors and staff to learn from their more detailed knowledge of areas and activities, and to get concerns and opinions about health and safety issues in the workplace; and
   • Looked at the accident book.
2. The manager then wrote down who could be harmed by the hazards and how.

3. For each hazard, the manager recorded what controls, if any, were in place to manage these hazards. She then compared these controls to the good practice guidance in HSE's Office health and safety web pages, Officewise and Essentials of health and safety at work. Where existing controls did not meet good practice, the manager wrote down what further actions were needed to manage the risk.

4. Putting the risk assessment into practice, the manager decided and recorded who was responsible for implementing the further actions and when they should be done. When each action was completed it was ticked off and the date recorded.

5. At an office meeting, the office manager discussed the findings with the staff and gave out copies of the risk assessment. The manager decided to review and update the assessment at least annually, or straightaway when major changes in the workplace occurred.

This example risk assessment is intended to show the kind of approach we expect a small business to take. It is not a generic risk assessment that you can just put your company name on and adopt wholesale without any thought. Doing that would not satisfy the law – and would not be effective in protecting people. Every business is different – you need to think through the hazards and controls required in your business.

Company Name: *ABC Office* **Date of Risk Assessment: 30-05-06** **REVIEW DATE: 28-05-06**

Source: HSE

What are the hazards?	Who might be harmed and how?	What are you already doing?	What further action is necessary?	Action by whom	Action by when	Done
Slips, trips and falls	All staff and visitors may suffer sprains or fractures if they trip over trailing cables/ rubbish or slip on spillages.	• Reasonable housekeeping standards maintained. • Cabinet drawers and doors kept closed when not in use. • Trailing cable from electrical machinery managed. • Floors, staircases and doors cleaned on a regular basis by the cleaners. • Stairs well lit and handrail provided. • Entrance well lit.	• Housekeeping to be discussed at regular staff meetings. • Supervisors given the responsibility of maintaining standards in their areas. • Office manager to carry out 3 monthly inspections to ensure adequate standards are maintained. • Instructions given that spillages should be cleaned up and dried immediately. • Repairs and maintenance carried out when necessary.	AB JC & OM AB JC & OM	26-06-06 5-6-2006 15-08-06 5-6-2006	26-06-06 5-6-2006 5-6-2006
Manual handling Deliveries: paper (regular) Office equipment (infrequent).	All staff (especially 'named staff') and staff of contract paper suppliers could suffer from back pain if they carry heavy/bulky objects in awkward places (e.g. staircases).	• Trolley used to transport boxes of paper, etc. • Only 'named staff' move office equipment (e.g. computers) and other heavy loads. • Top shelves used for storage of light boxes only.	• Need for manual handling training of named staff to be kept under review. • Supervisors to remind staff that heavy equipment to be moved by named staff only. • Agree, by contract, with paper suppliers of for delivery to point of store, (i.e. store cupboard).	AB JC & OM AB	11-7-2006 30-06-06 14-08-06	 27-06-06

What are the hazards?	Who might be harmed and how?	What are you already doing?	What further action is necessary?	Action by whom	Action by when	Done
Regular computer use	All office staff may suffer from upper limb disorders (RSI) from regular use of PCs or suffer headaches if lighting/picture is poor.	• Adjustable equipment, chair and footrest supplied. • Free eye test provided to all those working regularly with PCs by arrangement with local optician. • Venetian blinds provided to control ambient light. • All workers to carry out self-assessment from CD ROM within 6 weeks of starting/moving. • One member of staff complained of slight discomfort. Did not know how to adjust the equipment correctly.	• Supervisors to ensure staff know how to adjust equipment for own comfort. • Glasses to be provided to anyone working regularly with PCs where optician identifies they need them specifically for work with PC (not where just required for general use). • Action to be taken on the results of CD ROM self-assessment within 6 weeks. Individual results to be checked by appointed person and kept on file.	JC & OM JC & OM JC & OM	19-06-06 As required As required	12-6-2006
Stress	All staff could be affected by excessive pressure at work – from work demands, lack of job control, too little support from colleagues, not knowing their role, poor relationships, or badly managed change.	• Stress Policy in place. • Work plans and work objectives are discussed and agreed with staff each year.	• Team meeting held to discuss local causes of stress and develop some practical improvements; • Stress action plan aimed at tackling causes of stress agreed with staff; • Plan checked regularly to ensure it's being put into effect.	AB AB AB	26-06-06 10-7-2006 7-8-2006	26-06-06

What are the hazards?	Who might be harmed and how?	What are you already doing?	What further action is necessary?	Action by whom	Action by when	Done
Electrical	All staff could incur electrical shocks or burns if they use faulty electrical equipment.	• Sufficient sockets provided. • Staff trained to report defective plugs or cable to manager. • Photocopiers and computer systems maintained on contract. • Staff bringing in own kettles.	• 3 monthly visual inspection of electrical equipment to be carried out by office manager.	AB	15-08-06	
			• 2 yearly inspection and testing of portable heaters by local electrician.	AB	3-12-2007	
			• Staff instructed not to bring in their own kettle, as maintenance cannot be assured.	JC & OM	11-8-2006	
			• Water heater and coffee machine to be provided.	AB	11-8-2006	
Fire	If trapped in the office all staff and visitors could suffer from smoke inhalation or burns.	• Fire evacuation procedures displayed at each fire alarm point. • Fire drills twice yearly. • Exits and fire exits clearly marked. • Access to exits and extinguishers to be kept clear at all times. • Fire alarms maintained and tested by manufacturer. • Wastes bins emptied daily by cleaners.	• Fire extinguishers inspection to be put out to contract urgently.	AB	8-6-2006	6-6-2006
			• The office manager to make regular inspections to ensure that fire rules are followed and housekeeping standards are maintained.	AB	5-7-2006	
			• Training on use of extinguishers to be organised for identified staff.	AB	30-06-06	30-06-06

What are the hazards?	Who might be harmed and how?	What are you already doing?	What further action is necessary?	Action by whom	Action by when	Done
Bleach and strong detergents	Direct skin contact with could lead to the cleaner getting skin irritation. The vapour may cause eye irritation or breathing difficulties.	• None	• Cleaner to try safer alternative to bleach.	HF	24-07-06	
			• Information on correct use obtained from product instructions for use and data sheet. Cleaner to be made aware of these and what to do in case of splashing or spillage.	OM	7-6-2006	7-6-2006
			• Protective rubber gloves to be provided	OM	7-6-2006	5-6-2006
Smoking	Passive smoking can damage the health of all staff.	• 'No Smoking' policy adopted in the building. Smokers to go outside for a cigarette.	• Material on smoking cessation scheme obtained from local primary care trust and made available.	JC	28-06-06	
Hygiene and welfare	All staff could experience general discomfort.	• Toilets supplied with hot and cold water, soap and towels. • Washup area provided with drinking water and a fridge and cleaned daily.	• Office manager to monitor performance of cleaners.	AB	23-06-06	28-06-06
Environmental comfort factors.	All staff may feel too hot/cold or suffer other general discomfort.	• Building kept reasonably warm and light, window open to provide fresh air, plenty of space in offices. • No complaints from employees concerning personal comfort.	• No further action required.			

Example 2 – Small restaurant Example

General risk assessment

Company	
Location	
Assessor	
Date	Updated

Hazard	Who might be harmed?	Existing controls	Standard required	Action required
Fire and other emergencies	Employees visitors and customers	Fire alarm/Smoke stop doors Emergency lighting Fire extinguishers with annual maintenance contract Notices and Fire Drills Removal of waste packaging materials daily. Ash trays provided for smoking	Escape routes kept clear. Fire drills 2× per year. Combustible rubbish removed. Training in the use of fire extinguishers. Emergency lighting Clean filter on extractor hood in kitchen regularly.	Carry out fire drill
Slips trips and falls Access ways	Employees visitors and customers.	Regularly inspected, kept free of obstruction, well lit and good floor finishes maintained. Good hand rail on stairs.	Corridors kept free of obstruction. Good housekeeping and well lit. hand rails on stairs	
Manual handling Product deliveries and office supplies.	Employees	Heavy goods delivered to rear by kitchen. Level floor to carry in. Rear stair case used for kegs. Lifted by two people.	All heavier loads split or carried by team of two.	Manual handling assessment should be done for kegs.
Electrical Kitchen, Office, drinks making, ice and dish washing.	Employees and visitors	Sufficient sockets supplied. Equipment inspected annually. Staff encouraged to report defects. Fused multi-plugs used.	System maintained to recognised standards and inspected by a competent electrician.	Arrange inspection of portable equipment

Hazard	Who might be harmed?	Existing controls	Standard required	Action required
Bleach and strong detergents	Employees and cleaners	Kept with cleaning materials and label information followed. Hygiene specialist refill as necessary	Minimal exposure required by COSHH Regulations. Use safer substitute when possible and follow precautions on labels.	
Falling Objects Items stored in high places	Employees	Fairly light materials are stored in upper racks. Good quality steps used to reach high areas.	Make sure steps are kept in good condition	
Hand Knives Cuts from knives	Employees in Kitchen	Good quality sharp knives used. Training given on safe methods of cutting.	Use of good quality well maintained knives. Proper methods of cutting employed with good lighting and adequate space.	
Machinery Fridge, dishwasher and microwave	Employees and visitors	Regularly inspected by competent electrical engineer.	Kept in good condition and properly maintained through a competent person	
Gas installation for 4-burner and grill	Employees, visitors and customers	Installed by CORGI registered gas fitter. Inspected each year.	Properly installed and annual check by CORGI fitter.	
Hygiene and welfare	Employees, visitors and customers.	Clean male and female toilets, soap and hot water, air dryers provided. Separate washing facility in kitchen. First aid kept in the office. Specialist hygiene company contract.	Kept clean and germ free. Hot water and soap to be provided and means to dry hands.	
Environmental comfort Factors	Employees	Building kept reasonably warm and is well lit. Some opening windows and the door to the outside is often open. Extractor hood in kitchen and fan in restaurant.	Adequate heating lighting ventilation and space required.	

Hazard	Who might be harmed?	Existing controls	Standard required	Action required
Special Risks	There are no expectant or nursing mothers. One young person works occasional hours.	Rest breaks are arranged and young people do not work alone.	Follow the advice given in the HSE guidance	
Hot splashes and steam	Employees	Training of staff in the carrying of hot liquids. Safe procedures for cleaning and draining deep fryer. Good maintenance of coffee machine.	Equipment kept in good condition and proper training given. Experienced people to supervise.	

Example 3 General risk assessment – charity that has people with disabilities

Location	
Assessors	
Persons Seen	
Date (reviewed)	

Hazards	Who might be harmed	Residual risk H M L	Existing controls	Standard required	Action required
Fire and other emergencies	Employees, visitors, volunteers and Pw Disability	Medium	Fire certificated building changes to front lobby are now included in fire certificate to the satisfaction of the Fire Officer. Fire alarms/Smoke stop doors/Fire exit doors immediately and easily opened/Escape staircase at rear/ fire resistant enclosures for rear staircase/Fire detectors in certain areas/Emergency lighting, fitted and tested monthly. Magnetic catches linked to fire alarm fitted at rear lobby on fire doors. Fire extinguishers with annual maintenance contract. Alarms tested weekly. Notices displayed. Once yearly fire drills. Fire wardens appointed and trained.	Fire certificate requirements maintained. Fire Log Book to be kept. Escape routes kept clear. Fire drills 2× per year or as fire certificate. Fire alarm tested weekly. Emergency lighting checked monthly and tested as fire certificate Combustible rubbish removed. Emergency lighting tested annually by competent person. Training on the use of fire extinguishers for fire wardens. Training on emergency procedures for all.	1. Reception area being reinstated with fire doors returned to their original locations.

Hazards	Who might be harmed	Residual risk H M L	Existing controls	Standard required	Action required
			Removal of waste packaging materials daily. No smoking policy in building. Emergency chairs provided to assist persons with disability to escape down stairs.		
Gas appliances	Employees, visitors, volunteers and Pw Disability	Medium	Boilers installed and maintained by CORGI registered fitter. Hot pipes lagged and pipes marked.	Boilers installed and maintained by CORGI registered fitter. Annual inspection and test carried out. Pipes marked and lagged.	
Electrical Installation	Employees, visitors, volunteers and Pw Disability	Low	NICEIC test and inspection carried out by Mann Electrical Services on 4 December 1999. System 15 years old. New circuit added and actions required by survey carried out.	System maintained to recognised standards. NICEIC inspection and test carried out by a competent electrician every 5 years. RCD device fitted to system.	
Electrical equipment: Computers, printers, copiers, kitchen appliances.	Employees, visitors, volunteers and Pw Disability	Low	Equipment inspected by competent person and records maintained. Staff encouraged to report defects. Fused multi-plugs used. Photocopiers on contract. New furniture installed in 2001 with better layout and no cables across corridors, etc.	Sufficient sockets provided for all appliances. Equipment maintained properly. Cables properly controlled away from gangways. Portable Appliance Testing (PAT) every 1–2 years, as necessary. Visual checks by staff.	

Hazards	Who might be harmed	Residual risk H M L	Existing controls	Standard required	Action required
Slips trips and Falls Access ways Entrance steps	Employees, visitors, volunteers and Pw Disability.	Low	Regularly inspected, kept free of obstruction, well lit and good floor finishes maintained. Stairs kept in good condition with no lose treads.	Corridors kept free of obstruction. Good housekeeping and well lit. Floor finishes maintained properly.	
Falling Objects Racks, storage rooms	Employees	Medium	Strong stable racking used. Generally little used items at high level. Off site storage now arranged for financial papers.	Heavy loads to be stored at lower levels. Careful stacking necessary.	
Manual Handling Deliveries and office supplies	Employees and volunteers	Medium	Boxes limited to suitable weights. Team approach to moving heavy boxes, furniture or specialists used. Trolleys used for heavy goods and male staff required to help with lifting. Access kick stools provided in all stores. Strong access steps provided for higher storage rack access. ROSE reps received Manual handling training.	All heavier loads split or carried by team of two. Specialists movers used as necessary. Heavy items not stored at high level. No lifting of heavy loads above head height. Strong stable access steps available. Manual Handling Risk assessments carried out.	
Display screen Equipment Computers and work stations	Employees and volunteers	Low	Workstation risk assessments carried out for all employees who use equipment. Habitual Users under DSE Regulations are identified. Adjustable equipment and foot rests provided. Free eye tests if requested. Blinds and curtains used to control ambient light.	Suitable lighting, comfortable adjustable seats. For habitual users more detailed assessment of work stations required by the Health and Safety Display Screen Equipment Regulations	

Hazards	Who might be harmed	Residual risk H M L	Existing controls	Standard required	Action required
Hygiene and Welfare	Employees, visitors, volunteers and Pw Disability.	Low	Toilets supplied with hot and cold water. Contract cleaners used. Kitchen area supplied with drinking water, fridge, dishwasher, hot water heater and microwave oven. Kitchen regularly cleaned. First aid boxes kept on each floor. Trained first aider appointed.	Adequate toilet and washing facilities, rest area and a place to dry, change and store clothes. Adequate first aid boxes with notices. Trained first aider available at all times.	
Passenger lift	Employees, visitors, volunteers and Pw Disability. Lift engineer.	Medium	Otis, modern fully enclosed passenger lift. Monthly maintenance inspection by Otis, the manufacturers. Thorough examination every 6 months by Otis. Access to lift motor room for lift engineer only. Door kept locked and notice posted. Top of lift fitted with rails and controls for lift engineer's safety. Emergency lighting fitted.	Routine maintenance contract with competent person. Thorough examinations carried out in accordance with LOLER Regulations. Controlled access to lift motor room.	
Violence	Employees, visitors, volunteers and Pw Disability.	Low	Security arrangements with remotely controlled lock at main entrance. Key operated door latches into each office area.	Adequate access control maintained. Lone working only with special arrangements.	
Travelling in Company or others vehicles	Employees and other road users		Safe modern vehicles are used. Manufacturers service schedules are followed. Visual inspections carried out daily. Faults are rectified quickly. Mobile phones used	Modern vehicles used. Manufacturer service schedules are followed. Visual inspections carried out daily.	

Hazards	Who might be harmed	Residual risk H M L	Existing controls	Standard required	Action required
Environmental comfort Factors	Employees, visitors, volunteers and Pw Disability	Low	Building kept reasonably warm and is well lit. Ventilation via sufficient open windows throughout building. Space requirements are satisfactory. Provision for outside clothing provided. Special ventilation fitted in computer room.	Adequate heating lighting ventilation and space required. Provision for outside clothing to be stored during the day.	
Loan equipment cleaning Cleaning materials. Fleas from furniture. Dust and Aerosols from cleaning Hearing and eye risks	Employees, operator, visitors.	Medium	Plastic equipment cleaned with special wipes. Equipment returned to manufacturers for sterilisation when necessary. Upholstery work done by contractors offsite Small amount of stapling done using eye protection. Hearing protection (ear plugs) used for upholstery cleaning and stapling. PAT equipment calibrated and checked by manufacturer. High quality vacuum cleaner used with exhaust to the outside of the building. Properly maintained industrial upholstery cleaner used. Chemicals used according to labels and data sheets. They are not particularly hazardous materials. Periodic flea spray used. Noise levels checked – below first action level.	All equipment to be properly maintained and used in accordance with manufacturers instructions. Work to be carried out so that dust/aerosols do not get into the general air of the working area. Appropriate personal protective equipment to be used. COSHH, Noise and PPE assessments to be carried out. Effective control of fleas.	2. Carry out COSHH assessments on chemicals used.

Hazards	Who might be harmed	Residual risk H M L	Existing controls	Standard required	Action required
Asbestos	Building users	Low	As the building is only 15 years old there is no risk of asbestos containing materials being used. This has been confirmed by the building owner.	Positive knowledge of asbestos containing materials.	
Special Risks Expectant or nursing mothers. Young persons	Women of childbearing age work experience young people sometimes employed.		Rest breaks are arranged and young people do not work alone. Provision for pregnant or nursing mothers provided.	Follow the advice given in the H S E guidance for pregnant women or nursing mothers. Risk assessments must be sent to schools and parents if under school leaving age people are employed	
Stress	Staff	Low	A policy has been prepared and access to confidential counselling is provided.	Have a policy with guide lines. Give staff access to confidential counselling. Be sympathetic to employee needs.	

Managing health and safety

2

Summary

- **Managing Health and Safety**
- **The need to have a policy to plan, organise and monitor and safety issues**
- **The links with ISO 9000/2000, 14001 for the environment and safety management systems**
- **How to check your own plans for managing health and safety?**
- **What a simple safety policy looks like**

2.1 General management responsibilities

The guidance given in later chapters relates to specific risks or legal requirements. But there are some parts of the law that apply to practically all businesses. This is because employers and senior managers are the ones in control of the business. *They have the ability to prevent most accidents.*

Employers and senior managers have several key responsibilities, which are

1. to organize the work so that it is safe;
2. to appoint a person to provide health and safety assistance;
3. to provide adequate supervision;
4. to provide information, instruction and training; and
5. to monitor and review health and safety performance.

These are covered in more detail in the following paragraphs (see Fig. 2.1).

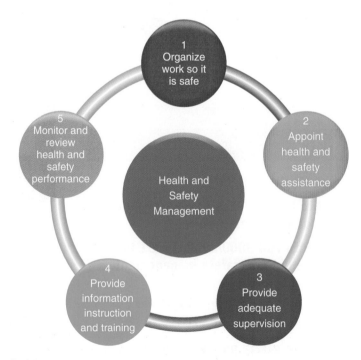

Figure 2.1 Health and safety management – essential elements.

2.2 Organize the work so that it is safe

Develop a *safety policy and strategy* for your business.

2.2.1 Safety policies

If you employ five or more people, by law you must prepare a written Safety Policy (*The Health and Safety at Work, etc. Act 1974*).

A Safety Policy is simply a document which describes *what steps you are taking to make sure that you are meeting your health and safety duties.* Having a Safety Policy will help you organise your health and safety at work so that you comply with the law.

The basic structure of the Safety Policy is laid down in the Health and Safety at Work Act. It consists of three main parts as shown in Fig. 2.2.

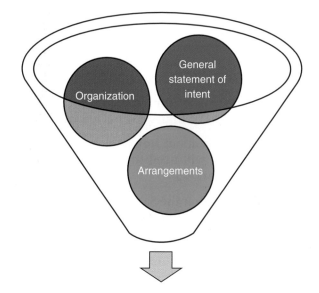

Figure 2.2 Elements of a Health and Safety Policy.

1. **General Statement of Intent**

   ◆ This is where you, as the employer make your *broad commitment to health and safety.*

   ◆ The person(s) in overall charge of the company should accept *ultimate responsibility* for health and safety matters. The statement should be *signed and dated* by the person accepting responsibility for the company.

Chapter 2

◆ Wording of the general statement is up to you, since it's **your** *commitment* and **your** company. An example is given in Appendix 2.2 to get you started.

2. **Organization**

◆ You need to make a list of health and safety *responsibilities* held by people in the company (i.e. people and their duties).

◆ You need to say *who* is responsible to *whom* and *for what*

◆ Most of the responsibilities for health and safety will be decided when the next part of the policy (*Arrangements*) is written.

◆ You will need to provide instructions on how to meet these responsibilities. The best place for this is in the Arrangements section of the policy

3. **Arrangements**

In this part you explain what the company needs to do to comply with the law and stop people being injured at work.

There are three main areas:

(a) **Legal requirements:** The specific health and safety laws and regulations which apply to your premises and line of work.

(b) **Hazards and risks:** Details of the hazards associated with your business activity and the risks of injury connected with those hazards.

(c) **Control methods:** This is about how you prevent or reduce the risks to acceptable levels. You will need to say what steps are being taken to make your place of work safe.

The most important thing is that the Safety Policy shows what is happening in a practical way. Here are some points to help you get started:

◆ You will need to describe what practical steps you are taking to make the workplace safe. Managers and Staff should be able to *refer to the Safety Policy* and find out what areas of health and safety they are responsible for and exactly how they are expected to meet those responsibilities.

◆ If there are *other* documents that relate to the Safety Policy, they should be *accurately referred* to and easy to find.

◆ You will need to make the safety policy part of your staff training. DO make sure that the right information is being given to the right people. DO NOT just give the Safety Policy to employees and expect them to read and understand it. They probably won't.

An example of a simple safety policy is given in Appendix 2.2.

2.3 Health and safety assistance

All businesses need help to comply with health and safety rules. The law requires a business to formally appoint someone to do this. In a small business it is likely that the owner can do the job themselves as long as they have the necessary knowledge.

If you feel that you have neither the time nor the expertise to deal with health and safety matters, you will need to appoint someone else to do the job. When you have an employee with the ability to do the job, it is better for them to be appointed rather than have someone from outside the business. An alternative is to use an external consultant.

You will need someone who knows and understands the work involved, who understands about assessing and preventing risk, who is up to date on health and safety and who can apply all this knowledge to the function.

2.4 Provide adequate supervision

The amount of supervision that any worker needs will depend on *the type of work, hazard and degree of risk involved* and *how much training and expertise they already have*.

To make sure that *rules are being followed* and necessary precautions are being taken, you yourself, as a person in charge of work activities, will need to make regular workplace checks (Fig. 2.3).

Figure 2.3 Adequate supervision to lift a canal boat.

If somebody disregards health and safety instructions, you MUST take action. Ignoring health and safety instructions is an illegal activity and if you ignore the situation it is no different from agreeing to it. *Individual managers may be personally liable for their own acts or omissions.*

2.5 Provide information, instruction and training

Anyone who is affected by what is happening in the workplace will need to be given safety information. This does not only apply to staff. It can also apply to visitors, members of public and if you have any, your contractors.

Information to be provided includes:

◆ who is at risk and why;
◆ how to carry out specific tasks safely;
◆ correct operation of equipment;
◆ emergency action;
◆ accident and hazard reporting procedures; and
◆ the safety responsibilities of individual people.

If you employ only a few staff, simple instructions and briefing sessions may be enough. But for larger companies a *formal in-house training programme* will be needed. It may be necessary to arrange for training to be provided by *external organizations* if you don't have the relevant expertise within the company.

There are many ways in which external training can be provided. Some of the most common are courses like 'Working Safely' (1 day) and 'Managing Safely' (5 days) accredited by the Institution of Occupational Safety and Health (IOSH) and run by many organizations (see Chapter 9 for contact details).

There are also Passport Schemes set up by some sectors or groups of companies to give workers basic health and safety awareness training. They are welcomed by the Health and Safety Executive (HSE), the Health and Safety Commission (HSC) and the Environment Agency, as they are a way of improving health and safety standards in the workplace. They also help promote good practice and can help reduce accidents and ill health caused by work. They are especially useful for workers and contractors who work in more than one industry or firm.

What are health, safety and environment Passports?

◆ A Passport shows that a worker has up-to-date basic health and safety or health, safety and environment awareness training. Some cover other subjects too.

◆ Passports are a way of controlling access to work sites – only workers with valid Passports are allowed to work.

◆ They are usually credit card size and made of strong plastic with a photograph and signature. Some have security features too, such as holograms.

◆ Workers can hold more than one Passport if they have been trained for work in more than one industry.

◆ They are a very simple way for workers who move from one industry to another, or work in more than one industry, to show employers that they have basic training.

◆ A Passport belongs to the worker not to the employer.

◆ Some Certification Schemes operate like Passports.

◆ Passports are a starting point for workers training for health, safety and environment qualifications.

What should Passport training cover?

A Passport holder should know about:

◆ the hazards and risks they may face;
◆ the hazards or risks they can cause for other people;
◆ how to identify relevant hazards and potential risks;
◆ how to assess what to do to eliminate the hazard and control the risk;
◆ how to take steps to control the risk to themselves and others;
◆ their safety and environmental responsibilities, and those of the people they work with;
◆ where to find any extra information they need to do their job safely; and
◆ how to follow a safe system of work.

Passports are not:

◆ a way of knowing or identifying that a worker is competent;
◆ a substitute for risk assessment;

Chapter 2

- ◆ a way of showing 'approval' of a contractor;
- ◆ required or regulated by law;
- ◆ a reason to ignore giving site-specific information; or
- ◆ a substitute for effective on-site management.

Keep a *record* of who has been trained; in what; by whom; and when.

Any safety signs or notices should comply with the Health and Safety (Safety Signs and Signals) Regulations 1996 (*see Chapter 3 for details*).

2.6 Monitor and review of health and safety performance

Are you able to identify hazards before they result in accidents? This must be a very important part of your strategy and there are established ways of achieving it:

- ◆ **Hazard reporting procedures:** These can be either formal or informal. Everyone in the workplace needs to understand that if they notice a hazard or a defect they must report it. People in charge of work activities should know how to take action when they receive these reports
- ◆ **Workplace inspections:** Make sure that you inspect your workplace at regular intervals that have been planned in advance. This will enable you to identify hazards that have not been picked up during a normal working day. See Appendix 2.3 for a simple inspection form which can be adapted for many workplaces (Fig. 2.4).

Figure 2.4 Inspection needed at this workplace.

- **Accident rates and investigation:** if there is an accident at work it should always be investigated to learn how it can be avoided in the future. Look at how often specific accidents happen, make sure that reporting is accurate and check whether any pattern is developing (see *Chapter 8: Accident and Emergencies*).

- **Health and safety audits:** It is good practice to have, every 2–3 years a more detailed audit of your safety systems and how you are complying with the law. To be really effective this should be carried out by a competent independent Health & Safety professional.

- **Safety policy review:** From time to time you will need to review the policy and revise it if necessary. If there have been any significant changes to the organisation of the business or to peoples' responsibilities, or the way in which work is carried out, the policy should be revised.

2.7 Major occupational health and safety management systems

For most small businesses this section will not be relevant, although you might like to read through it to familiarise yourself with the general idea. However some customers will require that you have the necessary accreditation to a recognised standard. For instance, if you are under contract to the National Health Service or a Social Services Department, you will probably need to know about and comply with one or more of these systems.

ISO stands for International Standards Office. Many people will have heard of the quality standards like ISO 9001/2000 which sets out a system of managing quality issues which can be checked by an external auditor. Large companies often require their suppliers to have ISO 9000/2000 accreditation to ensure that they have the systems in place to produce the products or services efficiently and consistently. There is also a similar standard for the environment ISO 14001.

OHSAS is the Occupational Health and Safety Assessment Series. OHSAS 18001 has been developed to be compatible with the ISO 9001 (Quality) and ISO 14001 (Environmental) management systems standards, in order to facilitate the integration of quality, environmental and Occupational Health and Safety (OH&S) management systems by organizations.

There is also a British Standard BS8800 which is essentially a guide to (OH&S) management systems.

It explains how the various elements in developing an (OH&S) management system can be tackled and integrated into day-to-day management arrangements, and how the system can be maintained as OH&S evolves, responding to internal and external influences.

The OHSAS 18001 standard is suitable for any business or organization that wishes to:

- Set up an OH&S management system to help it eliminate or minimise risk to their employees and other interested parties who may be exposed to OH&S risks associated with its activities.
- Implement, maintain and continually improve an OH&S management system.
- Assure itself that it is conforming to its stated OH&S policy.
- Show others that it is conforming to its stated OH&S policy.
- Seek certification/registration of its OH&S management system by an external organization.

In the UK, HSE has developed a similar system which is explained in their guidance HSG65. There is a lot of compatibility between OHSAS 18001 and HSG65. There is no legal requirement to follow these systems but they are useful guides and sometimes insurance companies will reduce premiums for Employers Liability insurance if a company follows a recognised system.

ILO-OSH 2001 was developed by the International Labour Organization (ILO) after an extensive study of many (OH&S) management systems used across the world. It was established as an international system following the publication of 'Guidelines on occupational safety and health management systems' in 2001. It is very similar to OHSAS 18001.

Appendix 2.1

Health and Safety – How do you comply?

Complete the assessment questionnaire to find out

For each **Yes** or **N/A** allow 1 point.   **No** or **D/K** gets zero points.

1. Health and safety policy

A written Health and Safety Policy is required if you have five or more people employed. There should be a brief general statement of intent.

		YES	NO	DK*	NA#
(a)	Do you have a written Policy?	☐	☐	☐	☐
(b)	Is it signed by a Director or equivalent?	☐	☐	☐	☐
(c)	Is it dated and reviewed each year?	☐	☐	☐	☐
(d)	Is a copy given to all employees?	☐	☐	☐	☐

* Don't Know

\# Not applicable

2. Organization

The Policy must include responsibilities. In many cases this will be done by position rather than by name.

		YES	NO	DK	NA
(a)	Are management responsibilities clearly set out?	☐	☐	☐	☐
(b)	Are responsibilities of employees set out?	☐	☐	☐	☐
(c)	Are responsibilities for specific issues, such as fire drills, safety inspections, accident investigations clearly set out?	☐	☐	☐	☐
(d)	Has someone been appointed to assist with implementing safety legislation?	☐	☐	☐	☐
(e)	Is the person(s) appointed competent in health and safety matters?	☐	☐	☐	☐

3. Health and safety arrangements

The health and safety arrangements should cover procedures and hazards at the work place. These should be kept in a file.

		YES	NO	DK	NA
(a)	Is there a comprehensive list of procedures for health and safety?	☐	☐	☐	☐
(b)	Are there written arrangements for dealing with the hazards of your business?	☐	☐	☐	☐
(c)	Do they cover safety, occupational health, emergencies and welfare?	☐	☐	☐	☐
(d)	Do you use a recognized safety management system like HSG65?	☐	☐	☐	☐

4. Consultation and information

You must consult employees on health and safety issues. The employer must provide adequate information. This can be done directly or through safety representatives.

		YES	NO	DK	NA
(a)	Do you have an established way of communicating with your staff about health and safety issues? This should include policy, objectives, performance, action and plans?	☐	☐	☐	☐
(b)	Have Union Safety Representatives or Representatives of Employee Safety (in non-union workplaces) been appointed?	☐	☐	☐	☐
(c)	Do you have a Safety Committee?	☐	☐	☐	☐

Chapter 2

5. Training

You must provide training for all employees to ensure they are familiar with health and safety issues generally, and especially with the procedures in your company.

	YES	NO	DK	NA
(a) Have health and safety training needs been identified?	☐	☐	☐	☐
(b) Have all members of management involved with people been given accredited training?	☐	☐	☐	☐
(c) Have all employees received adequate induction training on health and safety issues?	☐	☐	☐	☐
(d) Has specific training been given for first aid/fork trucks/manual handling, etc.?	☐	☐	☐	☐
(e) Do you keep all employees up to date on health and safety issues?	☐	☐	☐	☐

6. Risk assessments

General risk assessments, including fire risks, must be carried out and where more than 5 people are employed, you must record the significant findings. Other specific risk assessments must also be carried out.

	YES	NO	DK	NA
(a) Have general risk assessments been carried out including fire risks? (see Chapter 1)	☐	☐	☐	☐
(b) Are they up-to-date?	☐	☐	☐	☐
(c) Have COSHH assessments been carried out? (see Chapter 5)	☐	☐	☐	☐
(d) Have Manual Handling Assessments been carried out? (see Chapter 4)	☐	☐	☐	☐
(e) Have Display Screen Equipment (Computer Work stations) assessments been carried out? (see Chapter 6)	☐	☐	☐	☐
(f) Have noise assessments been carried out? (see Chapter 6)	☐	☐	☐	☐
(g) Have the special requirements of expectant mothers and young persons been assessed? (see Chapter 1)	☐	☐	☐	☐

Chapter 2

7. Monitoring health and safety performance

		YES	NO	DK	NA
(a)	Do you set performance targets for health and safety issues?	☐	☐	☐	☐
(b)	Do you have a procedure to monitor performance against these targets?	☐	☐	☐	☐
(c)	Do you count the number of accidents occurring?	☐	☐	☐	☐
(d)	Has the performance improved over the last 5 years?	☐	☐	☐	☐
(e)	Does the Board of Directors or equivalent consider health and safety performance?	☐	☐	☐	☐
(f)	Are health and safety audits carried out?	☐	☐	☐	☐
(g)	Are the audits independent?	☐	☐	☐	☐
(h)	Do Managers carry out safety inspections at least monthly?	☐	☐	☐	☐
(i)	Is there a system to implement audit and inspection actions points?	☐	☐	☐	☐
(j)	Does your company carry out a management review of health and safety activities?	☐	☐	☐	☐
(k)	Do you think health and safety issues are important in your company?	☐	☐	☐	☐

ASSESSMENT OF HEALTH AND SAFETY PERFORMANCE

For each **Yes** or **N/A** allow 1 point.   No or D/K gets zero points.

	Your Score	Maximum Score
1. Health and Safety Policy		4
2. Organization		5
3. Health and Safety Arrangements		4
4. Consultation and Information		3
5. Training		5
6. Risk Assessments		7
7. Monitoring		11
TOTAL:		39

Performance rating

Your Score

1–10   You need to seriously improve your commitment to health and safety. You are probably in breach of current UK health and safety legislation and could easily be prosecuted.

11–25   You have thought about health and safety management, but you do not have full commitment or achievement of excellence.

26–39   Your organization has a reasonable health and safety management system in place. However you must always be striving to improve the arrangements.

Appendix 2.2

Example of a simple health and safety policy

Example of a simple health and safety policy for a small company operating from a multi-storey office building.

General statement of intent

Health and Safety at Work Act 1974

This is the Health and Safety Policy of *Just do it Ltd*

The Directors of *Just do it Ltd* recognize and wholeheartedly accept their moral and legal obligations and responsibilities with regard to health and safety.

It is the Company's policy to take all reasonably practicable steps to:

- Provide adequate control of the health and safety risks arising from our work activities.

- Consult with our employees on matters affecting their health and safety.

- Provide and maintain safe plant and equipment.

- Ensure safe handling and use of substances.

- Provide information, instruction and supervision of employees.

- Ensure all employees are competent to do their tasks, and to give them adequate training.

- Prevent accidents and cases of work-related ill health.

- Maintain safe and health working conditions.

- Protect the health and safety of others who may be affected by our work activities.

- Review and revise this policy as necessary at regular intervals.

Signed _____

Date _____

Review Date _____

Health and safety organization

1. Health and Safety Responsibilities

The ultimate responsibility for safety and health is vested in the Directors of the Company who have the overall responsibility for the implementation of the Company Health and Safety Policy.

Day-to-day responsibility for ensuring that this policy is put into practice is delegated to Justin Doit, a Director of the Company.

To ensure health and safety standards are maintained and /or improved, the following people have specific responsibilities.

Name	Responsibility
Justin Doit	Safety inspections
	Incident investigations
	Policy and arrangements up-to-date
	Provision of safety information
Myway Doit	Product Design
	User operating instructions
Myway Doit	Safety training and instruction
Don Doit	First aid
Ree Doit	Maintenance of equipment

Employees have the responsibility to co-operate with managers to achieve a healthy and safe workplace and to take reasonable care of themselves and others.

- Each employee is responsible for following safe working practices, for taking a personal interest in promoting health and safety at work and for making a personal contribution to the achievement of high safety standards.
- Employees must comply with safety instructions applicable to their work and ask for advice from their immediate manager if in doubt on any safety matter.
- Employees have legal responsibilities and duties under the Health and Safety at Work, etc. Act 1974 and associated regulations including it.
- To take reasonable care for their own health and safety and that of any others who may be affected by their actions or omissions at work and to co-operate with the employer in meeting statutory requirements.

Chapter 2

- Not intentionally or recklessly to interfere with or misuse anything provided in the interests of health and safety at work.
- To use equipment in a safe manner and in accordance with instructions, and to report any defect in equipment which might compromise its safe use.
- To adhere to site rules when working on other employers' premises and not knowingly to place themselves at risk, reporting hazards and deficiencies.

Whenever an employee, supervisor or manager notices a health or safety problem, which they are not able to put right, they must straightaway tell the appropriate person named above.

Health and safe arrangements

1. Risk assessments

Risk assessments will be undertaken by Justin Doit who will be responsible to ensure that the significant findings are recorded and actioned.

The assessments will be reviewed annually or when the work activity changes, whichever is soonest.

2. Consultation with employees

The Company recognizes the duties imposed by the Health and Safety (Consultation with Employees) Regulations 1996. This gives rights to employees to be consulted and be provided with health and safety information.

All employees will be consulted directly on health and safety matters, through regular team meetings.

3. Safe plant and equipment

Ree Doit is responsible for identifying all equipment/plant needing maintenance. She will be responsible to ensure effective maintenance procedures are drawn up and implemented.

Ree Doit will check that new plant and equipment meets health and safety standards before it is purchased.

4. Safe handling and use of substances

Justin Doit will be responsible for identifying all substances, which need a COSHH (Control of Substances Hazardous to Health Regulations) assessment.

Justin Doit will be responsible for undertaking COSHH assessments and implementing actions as necessary. He will review the assessments annually or when changes are made whichever is soonest.

5. Information, instruction and supervision

The Health and Safety Law leaflets have been given to all employees and/or the Health and Safe law poster is displayed at the entrance.

Health and safety advice is available from our consultants Knowit All Ltd, telephone 01234 567890.

Supervision of young workers will be carried out by Myway Doit.

6. Competency for tasks and training

Training will be identified, arranged and monitored by Myway Doit.

7. Accidents, first aid and work-related ill health

The first-aid box is kept in the main office and the person appointed to look after first aid is Don Doit.

All accidents and cases of work-related ill health are to be recorded in the accident book. The book is kept by Don Doit.

Justin Doit is responsible for reporting accidents, diseases and dangerous occurrences to the enforcing authority, (for offices Local Authority). The requirements are contained in the Reporting of Injuries, Diseases and Dangerous Occurrences Regulations (RIDDOR) 1995.

8. Monitoring

To check our working conditions, and ensure our safe working practices are being followed we will carry out a safety inspection of the offices every 3 months.

Justin Doit will review our design instructions annually or when significant changes are made whichever is the soonest.

9. Emergency procedures – fire and evacuation

Ree Doit is responsible for ensuring that the fire risk assessment is undertaken and implemented.

Escape routes are checked daily by Myway Doit.

Fire extinguishers are maintained and checked by Pyro Fire Ltd.

Emergency evacuation will be checked every 6 months by Justin Doit.

10. Young Persons

A 'Young Person' is one who has not attained the age of 18. Companies are obliged to ensure that young persons at work are protected from any risks to their health or safety which are a consequence of:

- their lack of experience;
- their absence of awareness of existing or potential risks;
- the fact that young persons have not yet fully matured.

Young persons will be supervised by a competent person and the risks reduced to the lowest level that is reasonably practicable.

11. New and Expectant mothers

Once a woman has notified the Company that she is pregnant or is breast feeding, a risk assessment specifically for new and expectant mothers must be carried out taking into account the specific hazards which may effect new and expectant mothers.

Lone working by new or expectant mothers should not be allowed.

Pregnancy and breast feeding are normal parts of life and should not be associated with ill health. However whilst pregnant or breast feeding there MAY BE increased risks to the health of the mother and child from a range of hazards. The HSE has produced specific guidance in HS(G) 122, to help employers make the risk assessment.

12. Visitors

The host of any visitor to the Company's premises will take all reasonable practicable steps to secure their safety whilst on the premises.

13. Hazards

The hazards found in the workplace should be identified in the risk assessments with the necessary control measures.

Appendix 2.3

Example checklist for workplace audits (walk-through inspection)

Company _____ Location _____

Carried out by _____ Date _____

Item	OK	Not OK	Action
1. Are there any slip, trip or fall hazards, such as frayed carpets, trailing cables, wet floors or unprotected changes of floor level?			
2. Does the premises have a fire risk assessment?			
3. Is the fire plan displayed?			
4. Are fire extinguishers visible and accessible?			
5. Are the fire extinguishers suitable for the potential fire hazards and sufficient in number?			
6. Have the fire extinguishers been checked? (note the date of the previous Inspection)			
7. Are smoke detectors in place?			
8. Are the smoke detectors on scheduled maintenance?			
9. Are the names and locations of fire marshals and first aiders displayed and known to staff?			
10. Are all fire doors closed?			
11. Does the emergency lighting work?			
12. Is the HSE official poster 'Health and safety law – what you should know' displayed?			

Item	OK	Not OK	Action
13. Is the Employers Liability Insurance certificate displayed?			
14. Are there sufficient numbers of first-aid boxes?			
15. Do first-aid boxes have the correct contents?			
16. Is there a schedule for regularly checking the contents of first-aid boxes?			
17. Are eating facilities clean and adequate for the number of staff on site?			
18. Are toilet facilities clean and adequate for the number of staff on site?			
19. Are washing facilities clean and adequate for the number of staff on site?			
20. Are changing facilities clean and adequate for the number of staff on site?			
21. Is the working environment clean?			
22. Is the working environment at an appropriate temperature ?			
23. Is the working environment adequately lit?			
24. Is the working environment adequately ventilated?			
25. Is the working environment free from excessive noise and vibration?			
26. Are the guards in place on plant and machinery?			
27. Are individual items of plant and machinery labelled and identified?			
28. If there is plant and machinery do they have an appropriate service schedule?			

Item	OK	Not OK	Action
29. If chemicals are stored, are they stored appropriately in accordance with COSHH?			
30. Are stores/stockpiles safely stacked?			
31. Are routes free from obstructions?			
32. Are floors even and well maintained?			
33. Are there effective procedures to deal with spillages?			
34. Is waste stored appropriately and not allowed to accumulate?			
35. Are checks made on electrical appliances?			
36. Are company vehicles regularly serviced by a competent organization?			
37. Are employees wearing the correct personal protective clothing/equipment?			
38. Are correct manual handling techniques adopted?			
39. Are the workstations of display screen equipment users correctly laid out?			
40. Are users of display screen equipment provided with footrests, wrist rests, screen covers, etc. as necessary?			
41. Are lighting levels sufficient, that is, are 'dead' bulbs replaced?			
42. Are contractors competent to carry out their tasks – for example asbestos removal, or electrical system maintenance?			

Chapter 2

Framework of health, safety and fire law

3

Summary

- Some issues from the Health and Safety at Work Act 1974

- A few key general regulations for example:
  - Management
  - Consultation and safety representatives
  - Safety signs and notices

- Notification proceedings for new premises

- A health and safety checklist for starting a new business

- Workers rights under health and safety law

3.1 Legal framework

The wording in this chapter has to be quite formal because it comes from the law itself. We have simplified it as much as we can. Other chapters in the book explain things in more detail so where possible we have made reference to the chapters/sections to help you.

Because health and safety at work (HSW) is so important, there are rules, which require all of us not to put ourselves or others in danger. These rules are also there to protect the public from danger when they visit a place of work. The rules apply to all businesses, however small, to self-employed people and to employees.

There are two basic types of law (Fig. 3.1):

1. Criminal law: These are rules laid down by government which are imposed on its people for the protection of everyone. Breaking criminal law is an offence or crime and if prosecuted the person may get a fine or imprisonment. It has two areas or sources of law:
    (a) Common law – based on previous court judgements made by judges. Lower courts of law must follow the ruling of higher courts like the House of Lords.
    (b) Statute law – based on acts and regulations made by Parliament.
    Health and safety law is criminal law and people who breach the law are guilty of a crime. They would be prosecuted by an inspector or lawyer working for a government department.
2. Civil law concerns disputes between people and companies. A person sues another individual or company put right a civil wrong, normally by getting monetary compensation. In health and safety a person who is injured would sue a company or employer to get compensation for the injury. It also has the two areas or sources of law:
    (a) Common law.
    (b) Statute law.

Although there is a large amount of criminal health and safety law, there is no need for people at work to know it in detail. Instead, in the same way that the Highway Code regulates road users, people need to know what to do to comply with the law at work.

Health and safety law is aimed at preventing injury and ill health at work and making sure that adequate welfare facilities are provided. The legal framework is set out in the Health and Safety at Work Act 1974 (HSW Act) and in a number of regulations that have been issued under this Act.

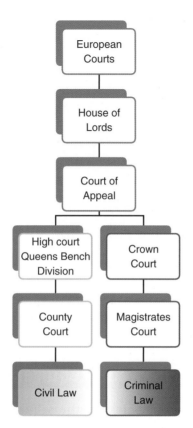

Figure 3.1 Framework for civil and criminal law in England and Wales.

These regulations give detailed guidance on particular aspects of health and safety such as management, hazardous substances, work equipment, work at height, noise and many other subjects (Fig. 3.2).

Most of the regulations made under the HSW Act apply to home-workers as well as to employees working at an employer's workplace. For this reason, employers must do a risk assessment of the work activities carried out by home-workers. More information is available in *'Home-working – Guidance for employers and employees on health and safety'* INDG226 (updated).

The fire safety legislation is now all contained within the Fire Safety Order of 2005 which does not come within the umbrella of the HSW Act and is therefore given a separate section in this chapter.

Chapter 3

Figure 3.2 Legal Framework – health, safety and fire statute law.

3.2 What the HSW Act requires

Under the HSW Act, employers and the self-employed are obliged to secure, so far as is reasonably practicable, the health, safety and welfare at work of anyone who may be affected by what the organization does, or fails to do. This includes all employees; trainees; part-timers; temporary staff; contractors; those who use the organization's workplace or equipment; visitors; employees working at home, the public who may be affected; people who use at work, products designed, made, supplied or imported by the organization.

This HSW Act applies to all work activities and to people in control of premises.

3.2.1 What basic actions must be taken?

To comply with the HSW Act and some of its basic regulations employers must:

◆ ensure the health, safety and welfare at work of all employees and others who may be affected, so far as is reasonably practicable;

* provide adequate training, information, instruction and supervision to employees;
* provide information and instruction to non-employees;
* provide and maintain safe plant and equipment;
* ensure safe handling, storage and use of substances;
* have a written, up-to-date health and safety policy (for five employees or more);
* display a current certificate as required by the Employers Liability (Compulsory Insurance) Act 1969 if anyone is employed;
* carry out risk assessments and, if five or more people are employed, the significant results must be recorded. The risk assessment includes special assessments for young people (under 18 years old) and new or nursing mothers, where they are employed;
* provide necessary personal protective equipment, free of charge, to employees;
* display the current health and safety law poster (ISBN 0-7176-2493-5) or hand out the equivalent leaflet (Fig. 3.3);
* notify, as specified (see Chapter 8 for more detail), certain types of injuries, occupational diseases and dangerous occurrences to the enforcing authorities or the central Incident Contact Centre (http://www.riddor.gov.uk or telephone 0845 300 9923);
* consult with employees either directly or through safety representatives on anything which might affect their health and safety. Consult with them about any information and training, which has to be provided (this may or may not be a trade union representative). Set up a safety committee if required by representatives;
* notify occupation of premises to your enforcing authority. (Forms: F9 Factories, F10 Construction, OSR1 Offices, Shops, etc. See 3.6, for more detail);
* not employ children of under school leaving age (if an industrial/factory business), apart from on authorised work experience schemes.

Chapter 3

3.2.2 Employee responsibilities

The employees' main responsibility is to co-operate with their employer in health and safety arrangements and to take all reasonable care of their own health and safety and that of others who may be affected by what they do.

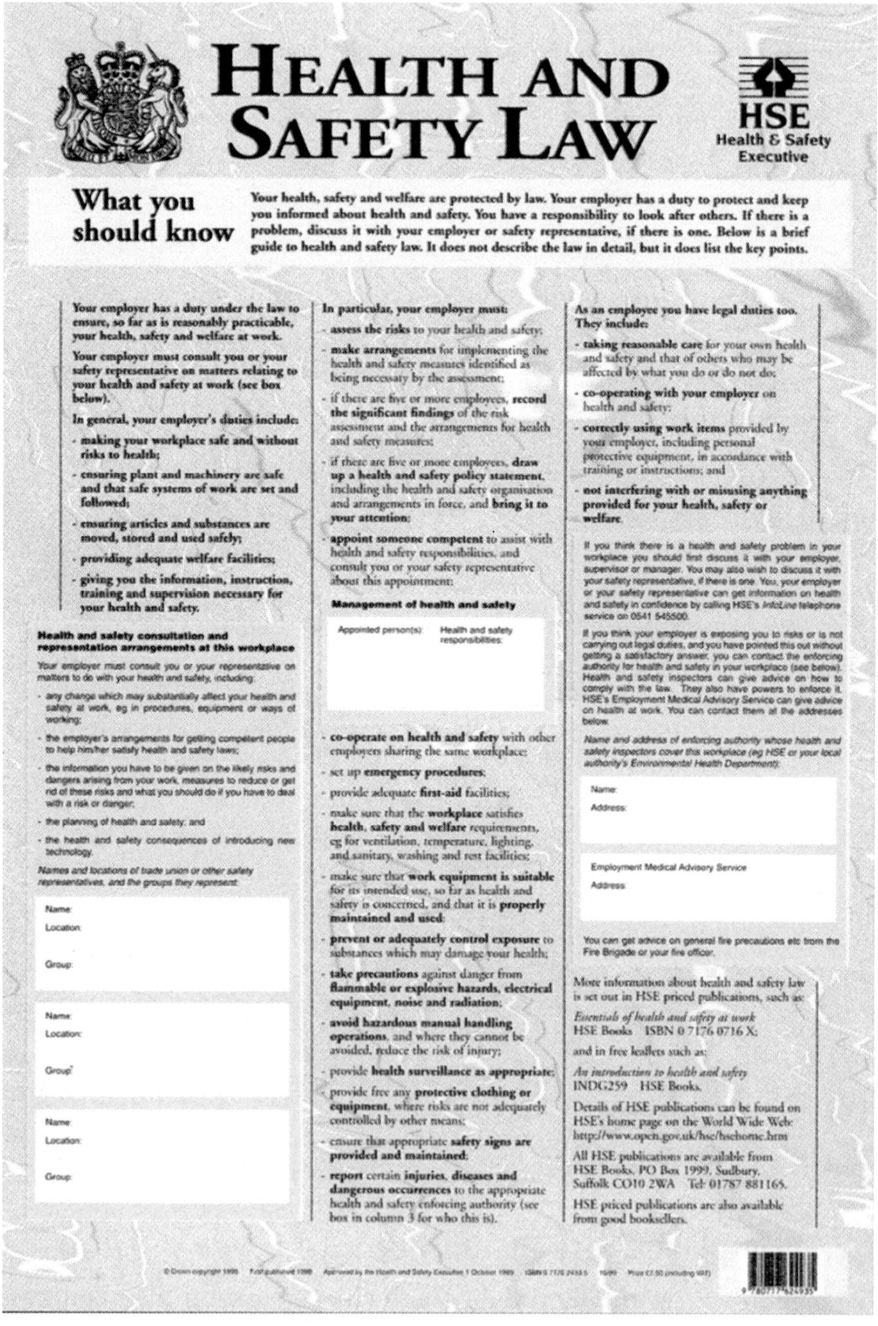

Chapter 3

Figure 3.3 Health and safety law poster – must be displayed or brochure given to employees.

This means that employees should:

- understand the hazards in their workplace and follow all safety rules and procedures;
- not behave in a stupid or reckless manner;
- use safety equipment and wear protective clothing when required to do so;
- not intentionally interfere with or misuse anything that has been provided in the interests of health and safety; and
- report hazards, near misses and accidents to a responsible person;

☞ *Stating Your Business* INDG324

☞ *Consulting Employees on Health and Safety* INDG232

Employees' rights as agreed with the HSE and the TUC are set out in Appendix 3.1.

3.3 Management Regulations

The Management of Health and Safety at Work Regulations 1999: These Regulations supplement the requirements of the HSW Act and specify a range of management issues, most of which must be carried out in all workplaces.

This is what you will need to know in order to comply with most of the Management Regulations:

1. *Risk assessment*: Every employer is required to make a 'suitable and sufficient' assessment of risks to employees, and risks to other people who might be affected by the business or organization, such as visiting contractors and members of the public. The risk assessments must take into account risks to new and expectant mothers and young people (see Chapter 1).

2. *Principles of prevention*: The following principles must be adopted when putting any preventative and protective measures into practice:
   - Avoid risks.
   - Weigh up the risks which cannot be avoided.
   - Combat the risks at source.
   - Adapt the work to the individual. Think about the design of the workplace, the choice of work equipment and the choice of working and production methods. Be careful about monotonous work and

work at a predetermined work-rate as these are especially likely to affect peoples' health and safety.

♦ Adapt to technical progress.

♦ Replace the dangerous by the non-dangerous or the less dangerous.

♦ Think about the whole of your business, the technology, organization of work, working conditions, social relationships and any other factors relating to the working environment, and develop a strategy that covers every aspect.

♦ Give collective protective measures priority over individual protective measures.

♦ Provide instructions and make sure that they are all understandable to employees.

3. *Effective arrangements for health and safety*: Procedures and arrangements must be devised (and recorded) for effective planning, organization, control, monitoring and review of safety measures (see Chapter 2).

4. *Health surveillance*: There may need to be health surveillance of staff, if for example you run a night shift or use dangerous chemicals. The Approved Code of Practice describes exactly what this is and when you are likely to need it.

5. *Competent assistance*: Every employer is obliged to appoint one or more 'competent person(s)' to advise and assist in doing what is necessary to comply with the law. They may be employees or outside consultants. The purpose is to make sure that all employers have access to health and safety expertise. It is better to give the job to an employee and they can be backed up by external expertise. You will need to make sure that the competent persons have adequate information, time and resources to do their job.

6. *Procedures for serious and imminent danger and contact with external services*: Procedures must be established for dealing with serious and imminent dangers, including fire evacuation plans and arrangements for other emergencies. There should be enough competent persons appointed to evacuate the premises quickly in the event of an emergency.

7. *Information for employees*: Information must be provided to staff on the risk assessment, risk controls, emergency procedures, the identity of the people appointed to assist on health and safety matters and any risks notified to the business by other people, for example suppliers of dangerous chemicals.

8. *Co-operation and co-ordination*: Where two or more employers share a workplace, each must co-operate with other employers in health and safety matters.

9. *Capabilities and training*: When giving tasks to employees, make sure they are capable of dealing with health and safety. They will need health and safety training:
   ◆ On recruitment.
   ◆ On being exposed to new or increased risks.
   ◆ On the introduction of new procedures, systems or technology.
   Training must be repeated periodically and take place in working hours (or while being paid).

10. *Duties on employees*: Equipment and materials must be used properly in accordance with instructions and training.

11. *New or expectant mothers*: Where work is of a kind that could present a risk to new or expectant mothers working there or their babies, the risk assessments must include an assessment of such risks.

12. *Young persons*: Employers must protect young persons at work from risks to their health and safety. Young people are more at risk because of their lack of experience. They may not be aware of existing or potential risks. The Regulations describe a variety of situations, which pose a significant risk to young people. They should not be employed in these except in the following situations. If they are over school leaving age they can be employed:
    ◆ When the work is necessary for their training.
    ◆ They will be supervised by a competent person.
    ◆ The risk will be reduced to the lowest level that is reasonably practicable.

3.4 Consultation and safety representatives

Consulting employees on health and safety matters can be very important to provide and maintain safe workplaces. By law you must consult your employees on health and safety matters. Consultation involves you in listening to and taking account of what your employees say before making any health and safety decisions. It is a lot more than just giving them information (Fig. 3.4).

You should consult employees about:

◆ any change which may substantially affect their health and safety at work;

◆ your arrangements for getting competent assistance to help you satisfy health and safety laws;

◆ the information that must be given about the risks and dangers which may arise from their work, and how they should be controlled;

Chapter 3

◆ the planning of health and safety training;

◆ the health and safety consequences of introducing new technology.

Figure 3.4 Consultation with employees is essential.

3.4.1 The Health and Safety (Consultation with Employees) Regulations 1996

Any employees not in groups covered by trade union safety representatives must be consulted by their employers under the HSCER 1996. The employer can choose to consult them directly or through elected representatives.

If the employer consults employees directly, he or she can choose whichever method suits everyone best. If the employer decides to consult his or her employees through an elected representative, then employees have to elect one or more people to represent them.

The Regulations apply to all employers and employees in Great Britain except:

◆ where employees are covered by safety representatives appointed by recognised trade unions under the Safety Representatives and Safety Committees Regulations (SRSCR) 1977;

◆ domestic staff employed in private households;

◆ crew of a ship under the direction of the master.

The basic requirements are:

1. *A duty to consult*: The employer must consult relevant employees in good time with regard to most health and safety issues and organizational matters.

2. *Who should be consulted*: Employers must consult with either the employees directly or someone (can be more than one) elected by employees in that group to represent them under these regulations. They are known as 'Representatives of Employee Safety' (ROES).

3. *Duty to provide information*: Employers must provide enough information to enable ROES to participate fully and carry out their functions.

4. *ROES*: Have the following functions (but no legal duties) on behalf of their group:
   (a) to make the employer aware of potential hazards and dangerous occurrences;
   (b) to make the employer aware of problems with the general health and safety of the group;
   (c) to represent the group of employees for which they are the ROES with health and safety inspectors.

A ROES can complain to an industrial tribunal that their employer has failed to permit time off for training or to be a candidate for election or their employer has failed to pay them.

3.4.2 The Safety Representatives and Safety Committees Regulations 1977

If you as an employer recognise a trade union and that trade union has appointed, or is about to appoint, safety representatives under the SRSCR 1977, then the employer must consult those safety representatives on matters affecting the group or groups of employees they represent. Members of these groups of employees may include people who are not members of that trade union.

If your consultation arrangements already satisfy the law then there is no need for change.

3.4.3 Protection of employees taking part in consultation?

The law protects employees against being dismissed or other action taken against them because they have taken part in health and safety consultation (whether as an individual or a representative). This includes taking part in electing a health and safety representative or being a candidate.

This section only applies to organizations that are unionised. Most small businesses are not. SRSCR 1977 prescribe the cases in which recognised trade unions may appoint safety representatives, specify the functions of such representatives require safety committees to be established and set out the obligations of employers towards them.

3.5 Safety signs and notices

Health and Safety (Safety Signs and Signals) Regulations 1996: The Regulations require employers to use and maintain a safety sign where there is a significant risk to health and safety that has not been avoided or controlled by other means, like engineering controls or safe systems of work, and where the use of a sign can help reduce the risk.

They apply to all workplaces and to all activities where people are employed, but exclude signs used in connection with transport or the supply and marketing of dangerous substances, products and equipment.

The Regulations require, where necessary, the use of road traffic signs in workplaces to regulate road traffic.

Functions of colours, shapes and symbols in safety signs
Safety colours
(a) Red

Red is a safety colour and must be used for any:

◆ Prohibition sign concerning dangerous behaviour (e.g. the safety colour on a 'No Smoking' sign). Prohibition signs must be round, with a black pictogram on a white background with red edging and a red diagonal line (top left, bottom right) (Fig. 3.5a and b).

◆ Danger alarm concerning stop, shutdown, emergency cutout devices, evacuate (e.g. the safety colour of an emergency stop button on equipment).

◆ Fire-fighting equipment.

Figure 3.5a Speed limit.   **Figure 3.5b** No passageway.

Red and white alternating stripes may be used for marking surface areas to show obstacles or dangerous locations.

(b) Yellow

Yellow (or amber) is a safety colour and must be used for any warning sign concerning the need to be careful, take precautions, examine or the like (e.g. the safety colour on hazard signs, such as for flammable material, electrical danger, etc.). Warning signs must be triangular, with a black pictogram on a yellow (or amber) background with black edging (Fig. 3.5c and d).

Figure 3.5c General warning.　　**Figure 3.5d** Forklift trucks.

Yellow and black alternating stripes may be used for marking surface areas to show obstacles or dangerous locations (Fig. 3.5e).

Figure 3.5e Striped obstacle marking.

Yellow may be used in continuous lines showing traffic routes.

(c) Blue

Blue is a safety colour and must be used for any mandatory sign requiring specific behaviour or action (e.g. the safety colour on a 'Safety Helmet Must Be Worn' sign or a 'Pedestrians Must Use This Route' sign). Mandatory signs must be round, with a white pictogram on a blue background (Fig. 3.5f and 5g).

Figure 3.5f Hearing protection.　　**Figure 3.5g** Gloves required.

(d) Green

Green is a safety colour and must be used for emergency escape signs (e.g. showing emergency doors, exits and routes) and first aid signs (e.g. showing location of first aid equipment and facilities). Escape and first aid signs must be rectangular or square, with a white pictogram on a green background. The green part must take up at least 50% of the area of the sign. So long as the green takes up at least 50% of the area, it is sometimes permitted to use a green pictogram on a white background, for example where there is a green wall and the reversal provides a more effective sign than one with a green background and white border; no danger (e.g. for 'return to normal') (Fig. 3.5h).

Figure 3.5h Means of escape sign.

3.6 Checklists for starting a new business

3.6.1 How do I get started on health and safety?

Controlling dangers at work is no different from tackling any other task – recognising the problem, knowing enough about it, deciding what to do and putting the solution into practice. Here are the 10 issues to get started:

1. Decide what could cause harm to people and how to take precautions. This is your *risk assessment.*
2. Decide how you are going to manage health and safety in your business. If you have five or more employees you need to write this down. This is your health and *safety policy.*
3. If you employ anyone you need *Employers Liability Compulsory Insurance* and you must display the certificate in your workplace.
4. You must provide free *health and safety training* in paid time, for your workers so they know what hazards and risks they may face and how to deal with them.
5. You must have *competent advice* to help you meet your health and safety duties. This can be workers from your business, external consultants/advisers or a combination of these. The Health and Safety Executive is an excellent resource.

6. You need to provide toilets, washing facilities and drinking water for all your employees, including those with disabilities. These are *basic health, safety and welfare needs*.

7. You must consult employees on health and safety matters.

8. If you have employees you must display the *health and safety law poster* (ISBN 0-7176-2493-5) or provide workers with a leaflet with the same information.

9. If you are an employer, self-employed or in control of work premises, by law you must *report some work-related accidents*, diseases and dangerous occurrences.

10. If you are a new business you will need to *register* either with the Health and Safety Executive (HSE) or your local authority – depending on the sort of business you have (see 3.6.3).

Chapter 3

3.6.2 Who enforces health and safety law?

Inspectors from the HSE or your local authority. For example: HSE at factories, farms and building sites; local authorities in offices, shops, hotels and catering, and leisure activities.

Inspectors visit workplaces to check that people are sticking to the rules. They investigate some accidents and complaints but mainly they help you to understand what you need to do. They enforce only when something is seriously wrong. The following leaflet from HSE Books provides more information: 'What to expect when a health and safety inspector calls'.

HSE operates a confidential telephone information service called InfoLine, which is open Monday to Friday between 8 a.m. and 6 p.m. You can contact InfoLine by telephone 0845 345 0055 or fax 02920 859260 or email hse.infoline@natbrit.com. Alternatively you can write to HSE Information Services, Caerphilly Business Park, Caerphilly CF83 3GG.

3.6.3 More about registering your business

If you are a new business you will need to register either with the HSE or your local authority – depending on the sort of business you have (see Fig. 3.6). Broadly speaking, you should register with:

- the HSE by completing form F9 (notice of factory occupation); or
- your local authority by completing form OSR1 (notice of employment of persons in office, shop or certain railway premises; this includes catering establishments, staff canteens, fuel storage depots).

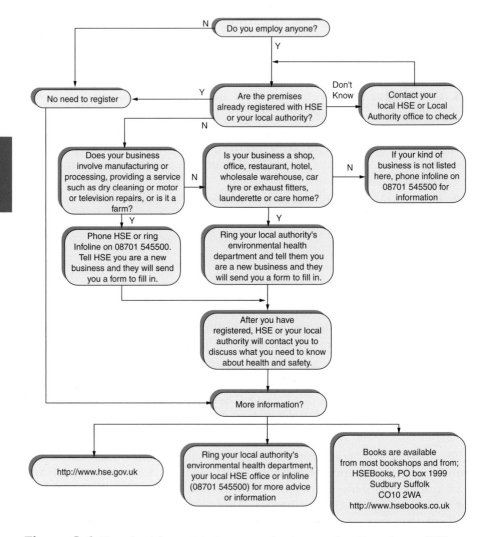

Figure 3.6 Flowchart for registering a new business or location. *Source*: HSE.

The definition of a factory is wide ranging, and covers most businesses where things are made, altered, adapted, repaired, decorated, finished, cleaned or demolished; or where people are employed in manual labour. Places where animals are slaughtered or held awaiting slaughter are also covered.

If you work with certain hazardous substances, such as asbestos or explosives, or in a hazardous industry such as construction or diving, you may also need to apply for a licence before your business starts to operate; or notify HSE or your local authority that you are starting certain specific activities. If you are already operating a business, you may also need to notify HSE or your local authority if you start certain specific activities.

Most farms will need to register but if in doubt please contact the HSE InfoLine who will be able to advise you (0845 345 0055).

If you are unsure whether you need to register or notify HSE or your local authority about the type of work you are doing, contact HSE's InfoLine on 0845 345 0055 or the Environmental Health Department of your local authority.

3.6.4 Corporate Manslaughter and Corporate Homicide Act 2007

Most small businesses will never be directly affected by this new Act which comes into force on 6 April 2008. However it is possible that larger organisations may impose stricter requirements on their contractors when they realize the gravity of the new offence.

The Act sets out a new offence for convicting an organisation where a gross failure in the way activities were managed or organised results in a person's death. The offence applies to all companies and other corporate bodies, operating in the UK, in the private, public and third sectors. It also applies to partnerships (and to trade unions and employers' associations) if they are an employer, as well as to Government departments and police forces. In England and Wales and Northern Ireland, the new offence will be called corporate manslaughter. It will be called corporate homicide in Scotland.

An organisation will be guilty of the new offence if the way in which its activities are **managed or organised** causes a death and amounts to a **gross breach** of a **duty of care** to the deceased.

Courts will look at management systems and practices across the organisation. Juries will consider how the fatal activity was managed or organised throughout the organisation, including any systems and processes for managing safety and how these were operated in practice. A substantial part of the failure within the organisation must have been at a **senior level. Senior level** means the people who make significant decisions about the organisation or substantial parts of it. This includes both centralised, headquarters functions as well as those in operational management roles.

New guidance, *Leading health and safety at work – Leadership Actions for Directors and Board Members*, is being drawn up jointly by the Institute of Directors and the Health and Safety Commission, and will be published UK-wide during 2008.

Appendix 3.1

Your health, your safety: A guide for workers

This information is from the HSE in collaboration with the Trades Union Congress (TUC). HSE is a government organization that works to protect the health, safety and welfare of workers by enforcing health and safety law and offering advice and support. The TUC represents over 70 trade unions with over 6.5 million members. It campaigns for fairness and decent standards at work.

If you are an employee (full- or part-time, temporary or permanent), this information explains what your rights are, what you should expect from your employer, what responsibilities you have and where to go for help. It also applies to you if you are a young person doing work experience, an apprentice, charity worker, mobile worker or home-worker.

If you are a temporary, casual or agency worker, the employment business/ agency, gangmaster, contractor or hirer you are working for has a legal duty to ensure you receive the rights set out here.

You have the right:

• To work in places where all the risks to your health and safety are properly controlled.
• To stop working and leave the area if you think you are in danger.
• To inform your employer about health and safety issues or concerns.
• To contact HSE or your local authority if you still have health and safety concerns and not get into trouble.
• To join a trade union and be a safety representative.
• To paid time off work for training if you are a safety representative.
• To a rest break of at least 20 minutes if you work more than 6 hours at a stretch and to an annual period of paid leave.

You must:

• Take care of your own health and safety and that of people who may be affected by what you do (or do not do). Co-operate with others on health and safety, and not interfere with, or misuse, anything provided for your health, safety or welfare.

Your employer must tell you:

• About risks to your health and safety from current or proposed working practices.
• About things or changes that may harm or affect your health and safety.

- How to do your job safely.
- What is done to protect your health and safety.
- How to get first aid treatment.
- What to do in an emergency.

Your employer must provide, free of charge:

- Training to do your job safely.
- Protection for you at work when necessary (such as clothing, shoes or boots, eye and ear protection, gloves, masks, etc.).
- Health checks if there is a danger of ill health because of your work.
- Regular health checks if you work nights and a check before you start.

(*Note*: If you are genuinely self-employed you are responsible for providing your own first aid arrangements, training, protective equipment and health checks, and for organising your own working time.)

Your employer must provide you with the following information: *Health and safety law: What you should know.* This should give the contact details of people who can help. Their health and safety policy statement. An up-to-date Employers' Liability (Compulsory Insurance) certificate visible in your place of work.

What to do if you are concerned about your health and safety: Phone HSE's InfoLine 0870-1 545-500 for advice or to complain, or the TUC's Know Your Rights line 0870 600 4882. If you would like to speak to someone in a language more suitable to you please call 08701 545500 and tell the operator which language. If you have lost your job because of a health and safety matter you may be able to complain to an Employment Tribunal. Ask your trade union or local Citizens Advice Bureau for advice.

HSE27(rev1) 11/04 *Published by the Health and Safety Executive.*

Source: TUC/HSE.

Chapter 3

Control of safety hazards

4

Summary

How to control safety hazards including:

- workplace, and basic welfare requirements
- movement of people, vehicles
- driving for work
- fire
- electricity
- work equipment
- manual handling
- slips and trips
- work at height
- confined spaces

4.1 The workplace and basic welfare

4.1.1 The Workplace Regulations

The Workplace (Health, Safety and Welfare) Regulations 1992 apply to most workplaces.

A place of work should be clean, comfortable and safe. Here is a summary of what you need to concentrate on to achieve this:

1. *Maintenance*: The workplace and equipment should be properly maintained (see Section 4.1.3).

2. *Ventilation*: There should be a sufficient quantity of fresh or purified air to ventilate enclosed workplaces.

3. *Temperature control*: If a room is used for more than short periods of time, then a comfortable temperature (normally 16C/60.8F) should be maintained during working hours. A thermometer should be available for staff to check it. Where uncomfortable conditions cannot be avoided, for example a cold store, access to a rest room should be made available.

4. *Lighting*: In all workplaces there should be good natural lighting, wherever possible. Emergency lighting should be provided where failure of normal lighting would cause danger. Lighting should be sufficient so that people can move around the building safely. In particular, staircases should be shadow free. Dazzling lights and glare should be avoided (Fig. 4.1).

Figure 4.1 Lighting at work.

5. *Cleanliness*: Workplaces (walls, floors ceilings) and furnishings should be kept clean (see Section 4.1.3).

6. *Space*: There should be enough floor area, height and unoccupied space in a workroom for people to feel comfortable. Allow at least 11 cubic metres per person (any height above 3 metres is discounted). When there is furniture in the room there must still be enough space left for people to move around easily.

7. *Well designed workstations*: These should be suitable for the worker and the work. You will need to provide good seating for any work that is to be done sitting down and there should be enough space around the workstation for people to work safely.

8. *Safe floors*: These should be suitable for their environment (see Sections 4.1.2 and 4.1.3).

9. *Danger of falling*: see Work at Height Regulations.

10. *Suitable windows*: Windows and glazed doors should be of safety material where danger could occur, that is, usually where the glazing is below shoulder level in doors and below waist level in windows. Large areas of glass that someone could walk into should be marked. Provision should be made so that window cleaning can be carried out safely for example, tilt and turn windows; fixing points for harnesses (also see Section 4.1.2).

11. *Safe traffic routes*: Safe routes should be organised for pedestrians and vehicles in workplaces (Fig. 4.2) (also see Sections 4.1.2 and 4.2.1).

Figure 4.2 Well marked passageway in large factory with crossing point.

12. *Safe escalators (and moving roadways)*: These must function safely. Safety devices and emergency stop controls should also be provided.

13. Good welfare facilities (see Section 4.1.4).

4.1.2 Access and egress

There should be a safe way to get to and from places of work and areas like toilets and washing areas. This will mean that floors, corridors and stairs will need to be free of obstructions, well lit with suitable properly maintained even surfaces and not slippery. There has to be space to get in and out safely, considering:

- vehicle movements;
- roadways, furniture or other equipment; and
- emissions from boilers or other process equipment;
- in bad weather outdoor routes need to be properly drained or salted/ sanded when icy and swept;
- where necessary there should be:
  - o safe glazing in doors and partitions suitably marked;
  - o handrails on open edges and floor openings; and
  - o good drainage in wet processes.
- Safe routes should be organised for pedestrians and vehicles and where practicable they should be kept separate.

4.1.3 Maintenance and cleanliness

The workplace and equipment should be maintained so that it works efficiently and is kept in good repair (for reasons of safety). Where necessary, there should be a system of maintenance for equipment and other devices. The system should provide records of work carried out and suitable forward dates for routine work. Examples of equipment would include emergency lighting, fire alarms, fences and anchorage points for safety harnesses. For things like fire extinguishers and gas boilers, maintenance contracts will be needed.

Poor housekeeping can create hazards through:

- dirty, unhygienic and untidy workplaces;
- hitting or tripping over obstructions in the workplace;
- blocking fire exit routes and doorways; and
- areas being too small and overcrowded.

Poor housekeeping can cause accidents and/or ill health and is a significant contributory factor in many incidents. It can be a factor in the spread of infection or the harm caused by the toxic effects of chemicals. Workplaces (walls, floors, ceilings) and furnishings should be kept clean and properly decorated as necessary. Waste materials should be stored in receptacles The standard of cleanliness should be appropriate to what is going on in the room (Fig. 4.3).

Figure 4.3 Poor housekeeping and waste collection.

Floors should be suitable for their environment especially if outside or in wet conditions. They should not be so uneven or slippery as to present risk. Arrangements should be made for the removal of spillage and obstructions as soon as possible

Regular inspections, high standards, effective cleaning and rubbish removal are all needed to ensure that poor housekeeping is prevented or rectified.

☞ *Preventing slips and trips at work* INDG225(Rev1)
☞ *Workplace health, safety and welfare* INDG244

4.1.4 Welfare

Basic welfare facilities required include:

- ◆ Clean well ventilated toilets which should, in most cases, be separate for male and female. For females means of disposing of sanitary dressings.
- ◆ Wash basins with hot and cold running water (Fig. 4.4).

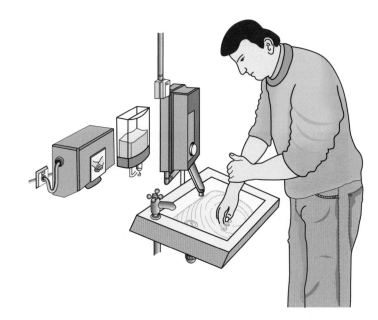

Figure 4.4 Washbasin with hot water, soap and dryer.

- ◆ Showers for dirty work.
- ◆ Soap and clean towels or hand dryer.
- ◆ Skin cleansers with nail brushes, barriers creams and skin condition-ing cream where necessary.
- ◆ Drying facilities for wet clothes with lockers or hanging space for clothing; Clothing accommodation is required where wet outdoor clothing can dry out during the working day.
- ◆ Changing facilities where special clothing is worn.
- ◆ Clean drinking water. A supply of drinking water with cups is required. Non-drinking water supplies should be clearly marked.
- ◆ Rest facilities including facilities for preparing and eating food. This should include food storage arrangements and safe methods of heat-ing food.
- ◆ Smoking is now forbidden in the workplace and in enclosed public spaces (see Chapter 5).
- ◆ Rest facilities for pregnant women and nursing mothers.

The number of facilities that must be provided are given in the INDG293 Welfare at Work as follows (Tables 4.1 and 4.2).

Table 4.1 Number of toilets and washbasins for mixed use (or women only)

Number of people at work	Number of toilets	Number of washbasins
1–5	1	1
6–25	2	2
26–50	3	3
51–75	4	4
76–100	5	5

Source: HSE

Table 4.2 Toilets used by men only

Number of men at work	Number of toilets	Number of urinals
1–15	1	1
16–30	2	1
31–45	2	2
46–60	3	2
61–75	3	3
76–90	4	3
91–100	4	4

4.2 Movement of people and vehicles

There are several ways in which vehicles are a hazard at work. These include being hit by a vehicle, causing injury or damage whilst driving, loading and unloading or refuelling.

Chapter 4

4.2.1 Precautions include

- ◆ keeping vehicles and people apart wherever possible by using separate entrances and barriers to make designated walkways (Figs. 4.5 and 4.6);
- ◆ setting up designated traffic routes, with appropriate signs and where appropriate one-way systems;

Figure 4.5 Shutting off a footpath to protect workers and the public.

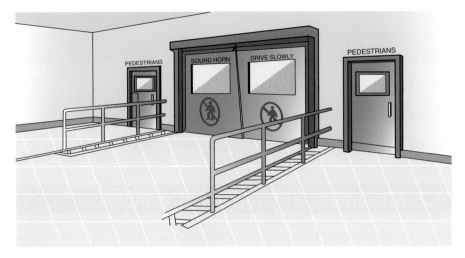

Figure 4.6 Separate doors for vehicles and pedestrians

- ensuring that drivers have good eyesight and their vision is not obscured by poor lighting, blind corners or large loads;
- minimising the need for reversing and including alarms and cctv on large vehicles;
- ensuring that pedestrians have good visibility with good lighting and mirrors on blind corners;
- ensuring that only trained and authorised drivers operate vehicles;
- not carrying passengers on vehicles unless there are properly fitted seats;
- ensuring that the correct vehicle is selected for the job and terrain;
- keeping vehicles properly maintained;
- only carrying out loading and unloading in suitable locations with appropriate equipment. Take care over access to vehicles and where forklift trucks are being used;
- taking proper precautions against fire when refuelling using flammable gases or liquids;
- having sensible speed limits;
- perhaps introducing speed reducing road humps;
- battery recharging gives off highly flammable hydrogen gases, which should be properly ventilated to the outside. This is particularly important when banks of forklift trucks are being re-charged in the same area.

☞ *Reversing Vehicles* INDG148

☞ *Managing vehicle safety at the workplace* INDG199

4.3 Driving for work

4.3.1 Introduction

Up to a third of all road traffic accidents involve somebody who is at work at the time. This may account for over 20 fatalities and 250 serious injuries every week. Some employers believe, incorrectly, that if they comply with certain road traffic law requirements, so that company vehicles have a valid MOT certificate, and drivers hold a valid licence, this is enough to ensure the safety of their employees, and others, when they are on the road. However, health and safety law applies to on-the-road work activities as it does to all work activities, and the risks should be managed effectively within a health and safety management system.

Health and safety law does not apply to commuting, unless the employee is travelling from their home to a location which is not their usual place of work.

Figure 4.7 A ferry port with drivers at work.

Early in 2007 a young driver of 23 who had been paralysed from the neck down when the light truck he was driving turned over was awarded £400,000 compensation which may rise to £1 million. He was driving at 22.00 hours having started at 03.30 hours, fitted two kitchens 122 miles apart and was returning to base with the Company's MD asleep beside him in the cab. This was considered to be a serious case of work related fatigue. It is likely that the MD would have been prosecuted for manslaughter had the driver died.

4.3.2 Evaluating the risks

The following considerations can be used to check on work-related road safety management

(a) **The driver**
  I. **Competency**
    ◆ Is the driver competent, experienced and capable of doing the work safely?

- Is the driver's licence valid for the type of vehicle to be driven?
- Is the vehicle suitable for the task or is it restricted by the driver's licence?
- Does recruitment procedure include appropriate pre-appointment checks?
- Is the driving licence checked for validity on recruitment and periodically thereafter?
- When the driver is at work, is he or she aware of company policy on work-related road safety?
- Are written instructions and guidance available?
- Has the company specified and monitored the standards of skill and expertise required for the circumstances for the job?

Figure 4.8 Rescue at a road traffic accident.

II. Training
Are drivers properly trained?
- Do drivers need additional training to carry out their duties safely?
- Does the company provide induction training for drivers?
- Are those drivers whose work exposes them to the highest risk given priority in training?
- Do drivers need to know how to carry out routine safety checks such as those on lights, tyres and wheel fixings?

◆ Do drivers know how to adjust safety equipment correctly, for example seat belts and head restraints?
◆ Are drivers able to use anti-lock brakes (ABS) properly?
◆ Do drivers have the expertise to ensure safe load distribution?
◆ If the vehicle breaks down, do drivers know what to do to ensure their own safety?
◆ Is there a handbook for drivers?
◆ Are drivers aware of the dangers of fatigue?
◆ Do drivers know the height of their vehicle, both laden and empty?

III. Fitness and health
◆ The driver's level of health and fitness should be sufficient for safe driving.
◆ Drivers of HGV's must have the appropriate medical certificate.
◆ Drivers who are most at risk, should also undergo regular medicals. Staff should not drive, or undertake other duties, while taking a course of medicine that might impair their judgement.

(b) The vehicle
Are vehicles maintained in a safe and fit condition?
There will need to be:
◆ maintenance arrangements to acceptable standards;
◆ basic safety checks for drivers;
◆ a method of ensuring that the vehicle does not exceed its maximum load weight;
◆ reliable methods to secure goods and equipment in transit;
◆ checks to make sure that safety equipment is in good working order;
◆ checks on seatbelts and head restraints. Are they fitted correctly and functioning properly?
◆ Drivers need to know what action to take if they consider their vehicle is unsafe.

(c) The journey
Has enough time been allowed to complete the driving job safely? A realistic schedule would take into account the type and condition of the road and allow the driver rest breaks. A non-vocational driver should not be expected to drive and work for longer than a professional driver. The recommendation of the Highway Code is for a 15 minute break every two hours.
◆ Are drivers put under pressure by the policy of the company? Are they encouraged to take unnecessary risks, for example exceeding safe speeds because of agreed arrival times?

◆ Is it possible for the driver to make an overnight stay? This may be preferable to having to complete a long road journey at the end of the working day?

◆ Are staff aware that working irregular hours can add to the dangers of driving? They need to be advised of the dangers of driving home from work when they are excessively tired. In such circumstances they may wish to consider an alternative, such as a taxi.

☞ *Managing Occupational Road Risk* – The RoSPA Guide, ISBN 9781850880530, www.rospa.com/morr

4.4 Fire

Fire kills from the effects of smoke and heat. It is one of the biggest single threats to any business. Whatever the nature of the business, you should be aware of how to prevent fire. Taking the right precautions will help to ensure the business doesn't go up in smoke.

4.4.1 The law

Regulatory Reform (Fire Safety) Order (RRFSO) 2005: This order, made under the Regulatory Reform Act 2001, reforms the law relating to fire safety in non-domestic premises. It replaces most fire safety legislation with one simple order. It means that any person who has some level of control in premises must take reasonable steps to reduce the risk from fire and make sure people can safely escape if there is a fire.

4.4.2 Where does the order apply?

The order applies to virtually all premises and covers nearly every type of building, structure and open space. For example, it applies to:

◆ offices and shops;

◆ premises that provide care, including care homes and hospitals;

◆ community halls, places of worship and other community premises;

◆ the shared areas of properties several households live in (housing laws may also apply);

◆ pubs, clubs and restaurants;

◆ schools and sports centres;

◆ tents and marquees;

- hotels and hostels; and
- factories and warehouses.

It does not apply to:

- people's private homes, including individual flats in a block or house.

4.4.3 What are the main rules under the order?

You must:

- carry out a fire-risk assessment identifying any possible dangers and risks;
- think about who may be especially at risk;
- get rid of or reduce the risk from fire and take precautions to deal with any possible fire risks left;
- make sure there is protection if flammable or explosive materials are used or stored;
- make a plan to deal with any emergency. Keep a record of your findings;
- review your findings if things change.

4.4.4 Who is responsible for meeting the order?

Under the order, anyone who has control of premises or anyone who has a degree of control over certain areas or systems may be a 'responsible person'. For example, it could be:

- the employer for those parts of premises staff may go to;
- the managing agent or owner for shared parts of premises or shared fire safety equipment such as fire-warning systems or sprinklers;
- the occupier, such as self-employed people or voluntary organisations if they have any control; or
- any other person who has some control over a part of the premises.

Although in many premises the responsible person will be obvious, there may be times when a number of people have some responsibility.

The RRFSO is normally enforced by the local Fire and Rescue Authority.

A series of guides has been produced and is available on www.firesafety-guides.communities.gov.uk

4.4.5 The fire triangle

In order for fire to occur, three ingredients need to be present:

- oxygen, present in the air;
- fuel such as soft furnishings, wood, paper and flammable liquids; and
- an ignition source, such as a naked flame, hot surfaces and electrical equipment

Take away any one of these and fire does not occur. The risk of fire is greatly reduced by basic good sense and good housekeeping. Here is some advice on common causes of fire and how to avoid them (Fig. 4.9).

Fuel
Flammable gases,
liquids, solids

Ignition source
Hot surfaces
Electrical equipment
Static electricity
Smoking materials
Naked flame

Oxygen
From the air
Oxidizing substances

Figure 4.9 Fire triangle.

4.4.6 Fire risk assessment

The responsible person must carry out a fire risk assessment and a record must be kept of significant findings (if five or more people are employed). It should pay particular attention to those at special risk, such as people with disabilities and those with special needs and must include consideration of any dangerous substance likely to be on the premises. The following five step approach is suggested by HM Government in 'A short guide to making your premises safe from fire' (see Fig. 4.10).

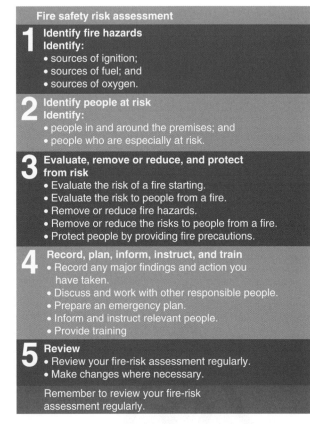

Figure 4.10 Five steps in fire risk assessment. *Source*: HM Government.

Step 1: Identify the hazards within your premises

You need to identify in line with the fire triangle:

- ◆ sources of ignition such as naked flames, heaters or some commercial processes;
- ◆ sources of fuel such as built-up waste, display materials, textiles or overstocked products; and
- ◆ sources of oxygen such as air conditioning or medicinal or commercial oxygen supplies.

Step 2: Identify people at risk

You will need to identify those people who may be especially at risk such as:

- ◆ people working near to fire dangers;

- people working alone or in isolated areas (such as in roof spaces or storerooms);
- children or parents with babies; and
- the elderly or infirm and people who are disabled.

Step 3: Evaluate, remove, reduce and protect from risk

Evaluate the level of risk in your premises. You should remove or reduce any fire hazards where possible and reduce any risks you have identified. For example, you should:

- replace highly flammable materials with less flammable ones;
- make sure you separate flammable materials from sources of ignition; and
- have a non\or safe-smoking policy (also see Chapter 5, Smokefree requirements).

When you have reduced the risk as far as possible, you must assess any risk that is left and decide whether there are any further measures you need to take to make sure you provide a reasonable level of fire safety.

You should follow the above guidelines with caution. You must look at each part of the premises and decide how quickly people would react to a warning of fire. If you are in any doubt or your premises provide care or sleeping facilities, you should read the more detailed guidance published by the Government or get expert advice. Some factories and warehouses can have longer travel distances to escape the fire.

Suitable fire exit doors

- You should be able to use fire exit doors and any doors on the escape routes without a key and without any specialist knowledge.
- In premises used by the public or large numbers of people, you may need push (panic) bars or push pads.

Other things to consider

- Whether you need emergency lighting.
- Suitable fire-safety signs in all but the smallest premises.
- Training for your staff or anyone else you may reasonably expect to help in a fire.
- A management system to make sure that you maintain your fire safety systems.

Chapter 4

Some very small and simple premises may be able to satisfy all these steps without difficulty. However, you should still be able to show that you have carried out all the steps.

Step 4: Record, plan, instruct, inform and train

In this step you should record, plan, instruct, inform and train. You will need to record the dangers and people you have identified as especially at risk in steps 1 and 2. You should also record what you did about it in step 3. A simple plan can help you achieve this (Figs. 4.11a and b).

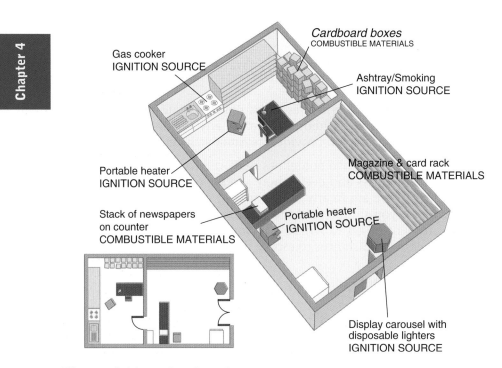

Figure 4.11a Before fire risk assessment.

You will also need to make an emergency plan, tailored to your premises. It should include the action that you need to take in a fire in your premises or any premises nearby. You will need to give staff, and occasionally others, such as hotel guests or volunteer stewards, instructions. All employees should receive enough information and training about the risks in the premises. Some, such as fire marshals, will need more thorough training (see Appendices 8.2 and 8.3 for examples).

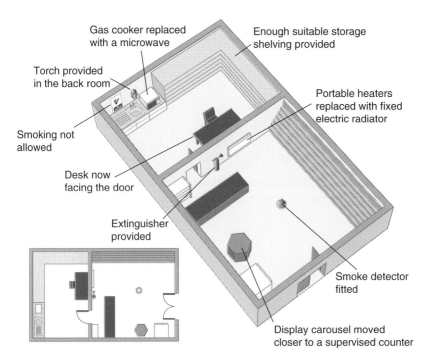

Gas cooker replaced with a microwave

Enough suitable storage shelving provided

Torch provided in the back room

Portable heaters replaced with fixed electric radiator

Smoking not allowed

Desk now facing the door

Extinguisher provided

Smoke detector fitted

Display carousel moved closer to a supervised counter

Figure 4.11b After fire risk assessment.

Step 5: Review

You should make sure your fire-risk assessment is up to date. You will need to re-examine your fire-risk assessment if you suspect it is no longer valid. For example, you may have a situation which could easily have resulted in a serious fire – often referred to as a 'near miss' – such as a fire in a waste paper bin in your office, or a napkin catching fire from a candle in a restaurant. You also need to re-examine your fire risk assessment every time there is a significant change to the level of risk in your premises. This could include:

- if you store more materials which can catch fire easily;
- a new night shift starting; or
- a change in the type or number of people using your premises.

4.4.7 Rubbish

Rubbish should be removed from the premises and into appropriate bins (with lids on) as quickly and as frequently as possible.

Never leave a fire, even in an incinerator, unattended. Make sure it is properly extinguished before it is left.

4.4.8 Smoking

A carelessly discarded cigarette end is still one of the most frequent fire starters. Getting rid of rubbish will help to reduce fires caused by smoking materials. Do not allow smoking outside in areas where goods or rubbish are stored, or in areas of high risk.

Since 1 July 2007 smoking has been banned inside buildings and many company vehicles (see Chapter 5).

4.4.9 Heaters

Heaters, especially portable ones, can be dangerous. All heaters should be sited well away from anything that can catch fire. Do not site them near walls and/or ceilings, which may be made of, or covered in, combustible materials. Portable heaters should be positioned so there is as little risk as possible of them being knocked over, pulled over or tripped over. Never stand books or papers on them or drape clothes over them – this may cause overheating which may result in fire.

4.4.10 Dangerous substances

If you use any dangerous chemicals, for example, paints or adhesives, keep them in a separate fire resistant storeroom well away from any source of heat. Aerosols, gas cartridges and cylinders can explode and start fires if they are exposed to heat. Always follow the manufacturer's instructions, paying special attention to storage recommendations (Fig. 4.12).

4.4.11 General precautions

There are a number of other good sense fire safety precautions you can take including the following:

◆ provide enough adequately signed exits for everyone to get out easily;

◆ provide fire escape doors, which can be opened easily from the inside whenever anyone is on the premises;

◆ never wedge fire doors open – they are there to stop smoke and flames from spreading;

Figure 4.12 Small flammable liquid store.

- ◆ if you have a fire alarm, check regularly that it is working. Can it be heard everywhere over normal background noise? Does everyone know what it is?
- ◆ provide enough properly serviced fire extinguishers of the right type, in order to deal promptly with any small outbreaks of fire;
- ◆ ensure that in accord with the Health and Safety (Safety Signs and Signals) Regulations, running person signs and direction arrows are correctly deployed. It is prudent to discuss this with the Local Authority Fire Officer. Ensure that panic bars or panic push pads on the fire exit doors are all correctly maintained and work efficiently;
- ◆ ensure everyone knows what to do in the event of a fire. Display clear instructions in popular places and have fire drills regularly: and
- ◆ ensure everyone knows how to raise the alarm and how to use the extinguishers.

4.4.12 Fire fighting equipment

Provide adequate means of fire fighting equipment. This is normally in the form of well-maintained fire extinguishers, blankets, etc. It is important to ensure that the correct fire extinguishers are provided for the hazards present in the premises.

Fire extinguishers are now coloured signal red with a colour-coded label or band. These replace the differently coloured metal cylinders previously in existence.

Note: Carbon dioxide and liquefied gases can be harmful if used in confined spaces. Ventilate the area as soon as the fire is extinguished (Fig. 4.13).

Blue		Standard dry powder or multi-purpose dry powder	For liquid and electrical fires. **DO NOT USE on metal fires**
Cream		AFFF (Aqueous Film Forming Foam)	A multipurpose extinguisher to be used on liquids or liquified solids
Cream		Foam	For use on liquid fires. **DO NOT USE on electrical or metal fires**
Red		Water	For wood, paper, textile and solid material fires. **DO NOT USE on liquid, electrical or metal fires**
Black		Carbon dioxide CO_2	For liquid and electrical fires. **DO NOT USE on metal**
Canary Yellow		Wet chemical	Cooking oils [Specially designed for high temperature cooking oils used in large industrial catering kitchens, restaurants and takeaway establishments etc.]

Figure 4.13 Guide to fire extinguishers and their use.

4.5 Electricity

4.5.1 Electrical hazards

The three main hazards of electricity are contact with live electrical parts, fire from overheating or overloading and explosion. Each year about 1000 accidents at work involving shock and burns are reported, and about 30 of these are fatal. Electricity passing through the body can cause convulsions (involuntary contractions of the muscles), the heart to stop beating and internal and external burns.

4.5.2 Ensuring that the electrical system is safe

Most of the electrical equipment used in offices and small businesses is low risk and does not need complex, frequent systems of testing but you do need to do some kind of formal inspection. It is usually enough to inspect it visually, concentrating on what you can see – damage or faults that you can put right. This way you will be able to prevent most electrical accidents from happening.

The sort of equipment we are talking about here is the type with a lead or cable and a plug; items which can be moved around such as photocopiers, fax machines and computers, kettles, heaters, fans, televisions, desk lamps, microwaves and vacuum cleaners. Extension leads and their sockets should also be inspected.

The most common problems are with leads and plugs, and sometimes with the equipment itself. The dangers are that people may suffer electric shock or that the equipment itself will cause a fire.

About 95% of electrical faults and damage can be discovered simply by looking. You should disconnect the piece of equipment from the power supply and then look for the following:

- damage to the cable covering (cuts, scratches, abrasions);
- damage to the plug (cracked casing, bent pins);
- non standard joints such as taped joints in the cable;
- overheating, shown by burn marks or stains;
- evidence that the equipment has been used in unsuitable conditions such as a place that is wet or very dusty;
- the coloured insulation of the internal wires showing;
- the outer covering of the cable not being gripped at the point where it enters the equipment or the plug;

◆ loose parts;
◆ loose screws;
◆ the outer casing of the equipment being damaged.

If the equipment has a moulded plug, you can only check the fuse, but for all other plugs, remove the cover and check:

◆ that a fuse is being used – not a piece of foil, wire or a nail;
◆ that there is no internal damage, or evidence that liquid has got into the plug;
◆ that there is no dirt or dust inside it;
◆ that the screws to the terminals are tight;
◆ that the wires are fitted to the correct terminals and none of the wires are bare except at the terminals (Fig. 4.14).

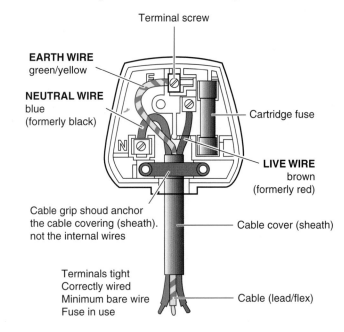

Figure 4.14 Standard UK 3 pin plug wiring.

To do these checks you will not need a qualified electrician. You can ask a member of staff to do them, provided that they know what to look for and how to look safely – for example by switching the appliance off and unplugging it before they start.

As to how often you should do these inspections, the HSE suggests that items that are used more frequently such as kettles, should be inspected more often and in their very helpful booklet 'Maintaining portable electrical equipment in offices and other low risk environments' INDG236 they provide a table suggesting suitable intervals for various types of equipment. Part of this table is shown in (Fig. 4.15).

Portable Appliance Testing

Offices and other low-risk environments only
Suggested <u>initial</u>* intervals

Equipment/environment	User checks	Formal visual inspection	Combined inspection and testing
Battery-operated: (less than 20 volts)	No	No	No
Extra low voltage: (less than 50 volts AC) eg telephone equipment, low voltage desk lights	No	No	No
Information technology: eg desktop computers, VDU screens	No	Yes, 2–4 years	No if double insulated – otherwise up to 5 years
Photo copiers,fax machines: NOT hand-held. Rarely moved	No	Yes, 2–4 years	No if double insulated – otherwise up to 5 years
Double insulated equipment: NOT hand-held. Moved occasionally, eg fans, table lamps, slide projectors	No	Yes, 2–4 years	No
Double insulated equipment: HAND-HELD eg some floor cleaners	Yes	Yes, 6 months–1 year	No
Earthed equipment (Class1): eg electric kettles, some floor cleaners	Yes	Yes, 6 months–1 year	Yes, 1–2 years
Cables (leads) and plugs connected to the above. Extension leads (mains voltage)	Yes	Yes, 6 months–4 years depending on the type it is connected to	Yes, 1–5 years depending on the type of equipment it is connected to

*NB: Experience of operating the maintenance system over a period of time, together with information on faults found, should be used to review the frequency of inspection.

It should also be used to review whether and how often equipment and associated leads and plugs should receive a combined inspection and test.

Figure 4.15 PAT testing table. *Source*: HSE.

The booklet also tells you where you can find guidance for businesses with a higher level of risk such as a small factory.

4.5.3 Test and examination

As we have said, most portable electric equipment used in small businesses is low risk and all you will need to do is visually inspect it. But it is a good idea to start off by doing a combined test (*Portable Appliance Testing or PAT) and inspection just to be sure that all your equipment is safe. Some types of fault cannot be seen just by looking, and this especially applies to earthing and the leads and plugs connected to earthed equipment. You can do the testing yourself, or train a member of staff, but you will need the right equipment and the expertise to use it and interpret the results. Even so, sometimes it may be necessary to seek advice from *outside expert electricians who will carry out a test and examination. This involves a more detailed investigation into the *installation and equipment.

*Installation: This is the fixed wiring from your meter through to socket outlets and light fittings. There is an accepted and recognised method for establishing whether your installation is in a safe condition and suitable for its actual use.

*Portable Appliance Testing: It is sometimes necessary to carry out PAT in which test equipment is used to check earthing, polarity, cable terminations and suitability for the environment. At this stage someone who is competent should set the frequency for the test examination.

*Outside contractors: When using an outside contractor, you must ensure that their technical knowledge or experience is appropriate for the work you are asking them to do, for example, membership of a recognised trade body such as the Electrical Contractors' Association (ECA) or the National Inspection Council for Electrical Installation Contracting (NICEIC) may be one indicator of competence. You must also be certain when employing someone with Electrical Joint Industries Board Qualifications that their expertise is in the relevant area, for example, a qualified television engineer may not be competent to rewire a building.

4.5.4 Keeping records

Although not a legal requirement, it is a good idea to keep written records. You will then know exactly where you are with inspections and tests and you can prove should you need to, that you have been complying with the law. For installation tests, you should obtain a certificate of compliance or test and examination such as that laid down in the current edition of the IEE Wiring Regulations. A log book should also be maintained for the results of the PAT testing.

4.5.5 Electric shock placard

All business need to display the electric shock placard in an area where electrical shock is possible like the electrical switch cupboard where the electrical consumer unit is housed (Fig. 4.16).

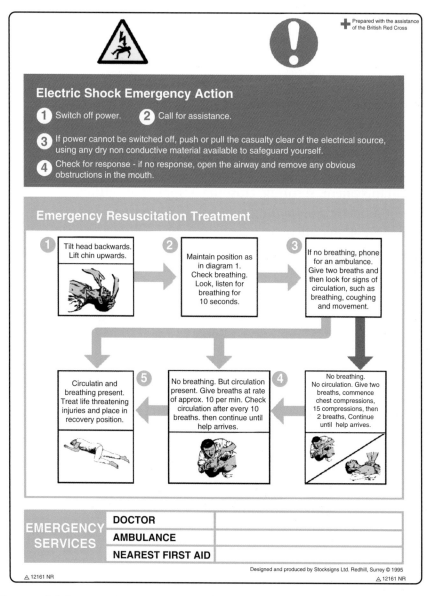

Figure 4.16 Typical electric shock action placard.

4.6 Work equipment

4.6.1 What are the risks?

Many serious accidents at work involve machinery – hair or clothing can become entangled in moving parts, people can be struck by moving parts of machinery, parts of the body can be drawn into or trapped in machinery, and/or parts of the machinery or work tool can be ejected.

Many things can increase the risks, including:

- ◆ not using the right equipment for the task, for example, ladders instead of access towers for an extended task at high level;
- ◆ not fitting adequate controls on machines, or fitting the wrong type of controls, so that equipment cannot be stopped quickly and safely, or it starts accidentally;
- ◆ not guarding machines properly, leading to accidents caused by entanglement, shearing, crushing, trapping or cutting;
- ◆ not properly maintaining guards and other safety devices;
- ◆ not providing the right information, instruction and training;
- ◆ not fitting roll-over protective structures (ROPS) and seat belts on mobile work equipment where there is a risk of roll over (excluding quad bikes) (Fig. 4.17);

Figure 4.17 Dump truck with ROPS (arch over driving seat) fitted.

- ◆ not maintaining work equipment or doing the regular inspections and thorough examinations; and
- ◆ not providing (free) adequate personal protective equipment (PPE) to use.

4.6.2 Risk identification

When identifying the risks, think about:

- the work being done during normal use of the equipment and also during setting-up, maintenance, cleaning and clearing blockages;
- which workers will use the equipment, including those who are inexperienced, have changed jobs or those who may have particular difficulties, for example, those with language problems or impaired hearing;
- people who may act stupidly or carelessly or make mistakes;
- guards or safety devices that are badly designed and difficult to use or are easy to defeat; and
- other things which could cause risks like vibration, electricity, wet or cold conditions.

4.6.3 Using machinery safely

Consider the following:

- Is the equipment suitable for the task?
- Are all the necessary safety devices fitted and in working order?
- Do you have the proper instructions for the equipment?
- Is the area around the machine safe and level with no obstructions?
- Has suitable lighting been provided?
- Has extraction ventilation been provided where required for example, on grinding and woodworking machinery?
- Has a risk assessments been done to establish a person's competence or training requirements to control particular machinery – very important for young people.
- Are machine operators trained and do they have enough information, instruction, training?
- Are people adequately supervised?
- Are safety instructions and procedures being used and followed?
- Are machine operators using appropriate everyday clothing without loose sleeves or open jackets or sandals?
- Have you, as the employer, supplied all necessary special PPE;
- Are safety guards or devices being used properly?

Chapter 4

◆ Is maintenance carried out correctly and in a safe way?

◆ Are hand tools being used correctly and properly maintained?

4.6.4 Guard dangerous parts of machines

Safeguarding the dangerous parts of machines and equipment that could cause injury is often the way in which risks are controlled. Remember:

◆ always use fixed guards wherever possible, securely fixed in position with bolts or screws which need tools to remove them (Fig. 4.18);

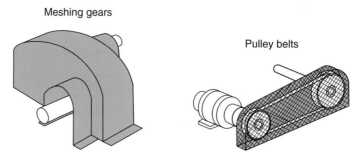

Figure 4.18 Fixed guards.

◆ use an interlocked guard for those parts which need regular access where a fixed guard would not be suitable. An interlocked guard must ensure that the machine cannot start before it is closed and the machine will stop if the guard is opened;

◆ safety devices like photoelectric systems or automatic guards may be used instead of fixed or interlocked guards on some machines like guillotines or presses;

◆ guards must be convenient to use and not easy to defeat;

◆ the materials used for the guards should be thought about carefully so that they continue to work correctly for example – plastic may be easy to see through, but can easily be scratched or damaged; wire mesh may have holes which allow access to the danger area; guards may also need to prevent harmful liquids, splashes or dust from escaping (Fig. 4.19);

◆ the guards must allow the machine to be cleaned and maintained safely; and

◆ where guards do not give proper protection use devices such as jigs, holders, push sticks.

Figure 4.19 Typical pedestal drill with guard.

Chapter 4

4.6.5 Safe use of mobile work equipment

Mobile work equipment carries out work while travelling or travels from one work place to another, for example dumpers, tractors, trailers and fork-lift trucks. Anyone riding on mobile work equipment needs protection from:

- ◆ falling out – fit cab guard rails, barriers (side, front or rear) or seat belts;
- ◆ unstable equipment – fit wider wheels or counterbalance weights to prevent the equipment rolling over. Fit ROPS (Fig. 4.17) and seat belts;
- ◆ falling objects. Fit falling object protective structures (FOPS). Fit a protective cage over the driver or have a strong cab.

Unless it is designed to do so do not let people ride on mobile work equipment. In exceptional circumstances it may be permitted, for example farm workers at harvest time riding on trailers. However you must have sides on trailers and/or secure handholds to prevent people falling.

4.6.6 Safe hand tools

Hand tools must be used and maintained properly (Fig. 4.20), for example:

- *Chisels*: keep edge sharpened to the correct angle. Use a wooden mallet on wood chisels and do not allow the head of cold chisels to spread to a mushroom shape – grind off the sides regularly;
- *Files*: these should have a proper handle and not be used as a lever;
- *Hammers*: avoid split, broken or loose handles and worn or chipped heads. Make sure the heads are securely attached to the shafts;
- *Screwdrivers*: never use them as chisels or use hammers on them – split handles can be dangerous;
- *Spanners*: use the correct size and avoid splayed jaws – discard when damaged. Do not use pipes, etc. as extension handles.

Figure 4.20 Typical handtools.

4.6.7 Maintenance

Equipment maintenance is crucially important. Poorly maintained equipment can be a safety risk, cause lost time and lead to a loss of quality of work produced. To prevent this:

- make sure the guards and other safety devices are checked and kept in working order;

◆ be alert for anyone disregarding or getting around the guards/safety devices; and

◆ check the safety devices after any modifications.

Inspections should be carried out by a competent person at regular intervals to make sure the equipment is safe to operate. The intervals between inspection will depend on the type of equipment, how often it is used and environmental conditions. Inspections should always be carried out before the equipment is used for the first time or after major repairs. Keep a record of inspections made as this can provide useful information for maintenance workers planning maintenance activities.

◆ Make sure the guards and other safety devices (e.g. photoelectric systems) are routinely checked and kept in working order. They should also be checked after any repairs or modifications by a competent person.

◆ Check the manufacturer's instructions for maintenance information to make sure it is done safely, where necessary and to the correct standard.

◆ Daily and weekly checks may be necessary, for example for fluid levels, pressures, brake function, guards. If you enter a contract to hire equipment, you will need to discuss what routine maintenance is needed and who will carry it out.

◆ Some equipment, for example cranes, excavators, need frequent preventive maintenance so that they remain safe to operate.

◆ Lifting equipment, pressure systems and power presses should be thoroughly examined by a competent person at regular intervals specified in law or according to an examination scheme drawn up by a competent person. Your insurance company may be able to give advice on who would be suitable to carry out this work.

4.6.8 Doing maintenance work safely

Many injuries are received when maintenance work is being done. Proper control includes the following safe working practices:

◆ where possible the power to the equipment should be switched off and ideally disconnected or have the fuses or keys removed, particularly where access to dangerous parts will be needed;

◆ isolate equipment and pipelines containing pressurised fluid, gas, steam or hazardous material. All Isolating valves should be locked off and the keys held securely by the person doing the work. The system should be depressurised where possible, particularly if access to dangerous parts will be needed;

♦ parts of equipment which could fall should be chocked;

♦ moving equipment should be allowed to stop;

♦ hot components should be given time to cool;

♦ switch off the engine of mobile equipment, put the gearbox in neutral, apply the brake and, where necessary, chock the wheels. Remove the ignition key or where none is fitted remove the battery connections;

♦ to prevent fire and explosions, thoroughly clean vessels that have contained flammable solids, liquids, gases or dusts and check them before hot work is carried out. Even small amounts of flammable material can give off enough vapour to create an explosive air mixture which could be ignited by a hand lamp or cutting/welding torch; this can even happen sometimes with a radio or power tool. Take expert advice on filling with water or inert gas before repairing fuel tanks or similar; and

♦ where working at height, make sure that a safe and secure means of access is provided which is suitable for the type, duration and frequency of the task.

☞ *Using work equipment safely* INDG229

4.7 Manual handling

4.7.1 Introduction

More than a third of all over-three day injuries reported to the HSE and local authorities arise from manual handling.

Manual handling involves lifting, lowering, pushing, pulling or carrying a load. Most harm from manual handling involves the back but there are also many cases of injuries to other parts of the body such as sprains to the legs or arms.

Employers should:

♦ avoid the need, as far as reasonably practicable for manual handling where there is a risk of injury;

♦ assess the risk of injury where potentially hazardous manual handling cannot be avoided (Fig. 4.21); and

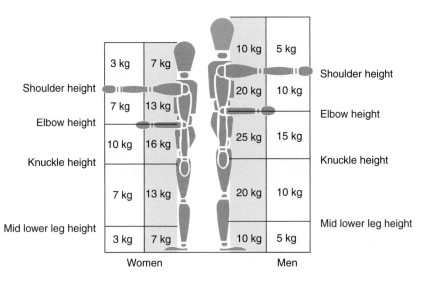

Figure 4.21 Guidance for manual lifting – recommended weights – HSE.

♦ reduce the risk of injury, as far as reasonable, by using mechanical aids, team working and adequate training.

4.7.2 Here are some questions to ask

Is it reasonably practicable to:

♦ Improve work place layout for example, improve the flow of materials or products?
♦ Reduce the amount of twisting and stooping?
♦ Avoid lifting from floor levels or above shoulder height?
♦ Cut carrying distances?
♦ Avoid repetitive handling?
♦ Vary the work, allowing one or more sets of muscles to rest while another is used?

Can the load be made:

♦ Lighter or less bulky?
♦ Easier to grasp?
♦ More stable?
♦ Less damaging to hold? that is free from sharp edges, rough surfaces, etc.

Is it possible to:

◆ Remove obstructions to allow more room for manoeuvre?

◆ Provide safer flooring that is free from spillages or slippery substances, etc?

◆ Avoid steps and steep ramps?

◆ Prevent extremes of hot and cold?

◆ Improve lighting?

◆ Consider less restrictive clothing or personal protective equipment?

Can the employer:

◆ Give employees more information, for example about the range of tasks they are likely to be asked to deal with?

◆ Provide comprehensive training (Fig. 4.22)?

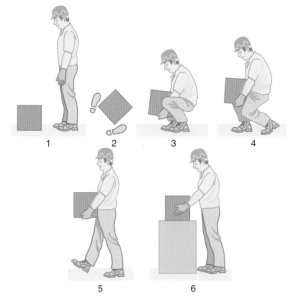

1. Check suitable clothing and assess load. Heaviest side to body.

2. Place feet apart – bend knees.

3. Firm grip – close to body slight bending of back, hips and knees at start

4. Lift smoothly to knee level and then waist level. No further bending of back.

5. With clear visibility move forward without twisting. Keep load close to the waist. Turn by moving feet. Keep head up do not look at load.

6. Set load down at waist level or to knee level and then floor.

Figure 4.22 The main elements of good lifting technique.

4.7.3 Manual Handling Operations Regulations

The Manual Handling Operations Regulations (MHO) 1992: apply to the manual handling (any transporting or supporting) of loads, that is by human effort, as opposed to mechanical handling by fork lift truck, crane, etc. Manual handling includes lifting, putting down, pushing, pulling, carrying or moving. The human effort may be applied directly to the load, or indirectly by pulling on a rope, chain or lever. Introducing mechanical assistance, like a hoist or sack truck, may reduce but not eliminate manual handling, since human effort is still required to move, steady or position the load (Fig. 4.23).

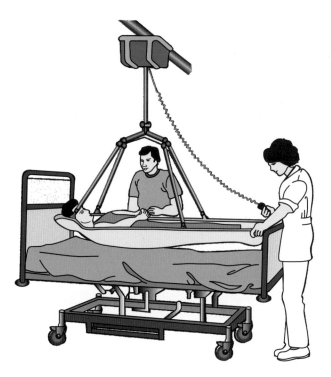

Figure 4.23 Mechanical aids to lift patients.

What is not included is the application of human effort for purposes other than transporting or supporting a load, for example, pulling on a rope to lash down a load or moving a machine control.

Injury in the context of these Regulations means to any part of the body. It should take account of the physical features of the load which might affect grip or cause direct injury, for example, slipperiness, sharp edges and extremes of temperature. It does not include injury caused by any toxic or corrosive substance which has leaked from a load, is on its surface or is part of the load.

The Regulations require employers to:

♦ avoid the need for hazardous manual handling, so far as is reasonably practicable (Fig. 4.23);

♦ assess the risk of injury from any hazardous manual handling that can't be avoided (Fig. 4.21); and

♦ reduce the risk of injury from hazardous manual handling, so far as is reasonably practicable (Fig. 4.22).

Employees have duties too. They should:

♦ follow appropriate systems of work laid down for their safety;

♦ make proper use of equipment provided for their safety;

♦ cooperate with their employer on health and safety matters;

♦ inform the employer if they identify hazardous handling activities; and

♦ take care to ensure that their activities do not put others at risk.

☞ *Getting to grips with manual handling* INDG143(Rev2)

☞ *Aching arms (or RSI) in small businesses: is ill health due to upper limb disorders a problem in your workplace?* INDG171(Rev1)

4.8 Slips and trips

4.8.1 Introduction

Slips and trips are the single most common cause of injuries at work. They account for over a third of all major injuries reported each year. These cost employers over £300 million a year in lost production, investigations, hiring alternative staff and other costs. Slips and trips account for over half of all reported injuries to members of the public.

Compensation claims brought as a result of an injury can be extremely expensive, time consuming and therefore damaging to business, especially where the public is involved. Insurance covers only a small proportion of the total costs.

Everyone at work, but particularly employers, can help to reduce slip and trip hazards through good health and safety management and practices. You will find that good solutions are simple, cheap and often give other benefits as well (Fig. 4.24).

Figure 4.24 Clean carefully to prevent slips.

4.8.2 Working practices

When dealing with slip and trip risks it is much easier if working conditions and practices are good from the start. Use suitable non-slip and even floor surfaces, ensure lighting levels are adequate, plan pedestrian and where practical separate traffic routes and avoid having too much clutter, work in progress and overcrowding (Table 4.3).

(a) Cleaning and maintenance

You must ensure that equipment used and your method of cleaning are suitable for the type, location and use of the surface being treated. You may need to get advice on this from the manufacturer or supplier of the equipment or cleaning materials. There may be sensible advice provided by the suppliers – ask if you do not have it to hand. Take care not to create additional slip or trip hazards while cleaning and maintenance work is being done by for example warning signs and temporary closures of some areas.

Carry out all necessary maintenance work promptly particularly where there are significant damaged areas. You may need to get outside help or guidance from flooring manufacturers or specialists. If necessary signs and barriers should be put up to keep people away from dangerous areas. Keep records so that the maintenance system can be checked.

Train workers in the correct use of any safety and cleaning equipment provided.

Table 4.3 There are many simple steps that can be taken to reduce risks. Here are a few examples taken from the HSE's leaflet INDG225

Hazard	Suggested action
Spillage of wet and dry substances	Clean spills up immediately. If a liquid is greasy ensure a suitable cleaning agent is used. After cleaning the floor may be wet for some time. Use appropriate signs to tell people the floor is still wet and arrange alternative bypass routes.
Trailing cables	Position equipment to avoid cables crossing pedestrian routes, use cable covers to securely fix to surfaces, restrict access to prevent contact.
Miscellaneous rubbish, for example plastic bags	Keep areas clear, remove rubbish and do not allow to build up.
Rugs/mats	Ensure mats are securely fixed and do not have curling edges.
Slippery surfaces	Assess the cause and treat accordingly, for example treat chemically, appropriate cleaning method etc.
Change from wet to dry floor surface	Suitable footwear, warn of risks by using signs, locate doormats where these changes are likely.
Poor lighting	Improve lighting levels and placement of light fittings to ensure more even lighting of all floor areas.
Changes of level	Improve lighting, add easily seen tread nosings.
Slopes	Improve visibility, provide hand rails, use floor markings.
Smoke/steam obscuring view	Eliminate or control by redirecting it away from risk areas; improve ventilation and warn of it.
Unsuitable footwear	Ensure workers choose suitable footwear, particularly with the correct type of sole. If the type of work requires special protective footwear the employer is required by law to provide it free of charge.

Source: HSE

Chapter 4

(b) Lighting

Good lighting should enable people to see obstructions and potentially slippery areas while not creating glare or dazzling people. Replace, repair or clean lights before levels become too low for safe work. In some places such as staircases or where levels change it is important to have local lighting. Variations in texture or colour can often help to improve vision.

(c) Floors

Badly maintained floors are a major cause of slips and trips. Floors need to be checked regularly for deposits of slippery materials, loose finishes, broken surfaces and holes, worn rugs and mats. Take care in the choice of floor if it is likely to become wet or dusty due to work processes.

(d) Obstructions

Poor housekeeping resulting in equipment, materials, and waste left lying around can easily go unnoticed and cause a trip. Keep work areas tidy and where obstructions can't be removed, use warning signs and/or barriers.

(e) Footwear

Good quality slip resistant footwear can play an important part in preventing slips and trips. The type of flooring material and whether it is frequently wet, will dictate the type of sole material to use. For example microcellular urethane and rubber are far more slip resistant than PVC or leather. Employers need to provide footwear free if it is necessary to protect the safety of workers.

4.9 Working at height (WAH)

Every year about 70 people die and nearly 4000 are seriously injured as a result of a fall from height while at work. Falls from height are the most common cause of fatal accidents at work.

Many activities involve working at height. These include: Ladders, step-ladders and trestles, scaffolding, MEWPS (Mobile elevating work platforms), prefabricated scaffold towers, roof work, work in lift shafts, sheeting vehicles, working where people can fall into pits and excavation.

4.9.1 Work at Height Regulations 2005 (WAH)

The Regulations apply to all WAH where there is a risk of a fall likely to cause personal injury. They place duties on employers, the self-employed, and any person who controls the work of others (e.g. facilities managers or building owners who may contract others to WAH) to the extent they control the work.

The Regulations do not apply to the provision of paid instruction or leadership in caving or climbing by way of sport, recreation, team building or similar activities.

They define 'WAH' as:

(a) work in any place, including a place at or below ground level; and

(b) obtaining access to or egress from such a place while at work, except by a staircase in a permanent workplace;

where a person could fall a distance liable to cause personal injury. This includes falls from relatively low platforms. The WAH Regulations also cover the possibility of material and other objects falling onto people below.

'Work' includes moving around at a place of work (except by a staircase in a permanent workplace) but not travel to or from a place of work. For instance, a sales assistant on a stepladder would be working at height, but it would not be likely that the Regulations would be applied to a mounted police officer on patrol.

The Regulations require a risk assessment for all work conducted at height and arrangements to be put in place to make sure that:

- all WAH is properly planned and organised;
- all WAH takes account of weather conditions that could endanger health and safety;
- those involved in WAH are trained and competent;
- the place where WAH is done is safe;
- equipment for WAH is appropriately inspected;
- the risks from fragile surfaces are properly controlled; and
- the risks from falling objects are properly controlled.

The risk assessment should be considered in the following order:

(a) *Avoid*: working at height where possible.

(b) *Prevent falls of people and objects*: This might include doing the work safely from an existing work place or choosing the right work equipment to prevent falls.

(c) *Mitigate*: the consequences of a fall by minimising the distance and choosing suitable fall arrest equipment.

(d) *Give collective*: protective measures (e.g. guardrails, nets, airbags) precedence over personal protective measures like a safety harness.

4.9.2 Scaffolds

Scaffolds provide the safest work platforms and must be erected and dismantled by competent specialists. They require inspections before use and every 7 days thereafter when in use.

4.9.3 Ladders

◆ Only use ladders for low risk tasks of short duration, that is less than 6 metres high, light work, no significant sideways or outward forces, secure handhold maintained.

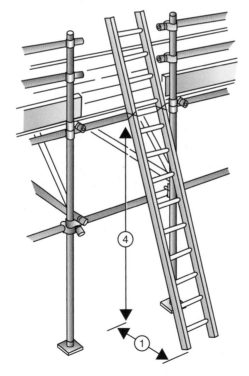

Figure 4.25 Correct angle for ladder.

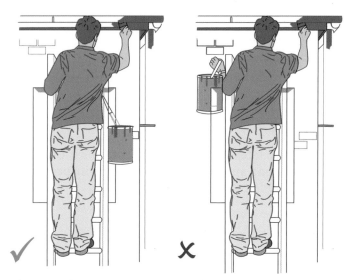

Attach paint cans and the like to the ladder

Pole ladder in use

Figure 4.26 Use of ladders – maintain three point contact.

◆ Use a safer alternative when possible.

◆ Place on a firm solid horizontal base.

◆ Lean at the correct angle −1 metre out for every four metres up (rungs are about 1/3 metre apart) (Fig. 4.25).

- All ladders should be secured from falling. This is done by tying at the top.
- Always maintain '3 point contact' when ascending or descending i.e. two hands one foot or two feet one hand (Fig. 4.26).

4.9.4 Stepladders

- Use stepladders on a firm horizontal base.
- Ensure they are long enough for the task.
- Open the legs out fully.
- Use front-on without any side loading to upset their stability.
- They should be fitted with a top loop handrail.

Do not stand on the small top platform. Platforms should only be used if they are designed for a work platform with a hand rail, such as on a metal wheeled stepladder (Fig. 4.27).

✗ Wrong way

Right way ✓

✗ Stepladder too short
✗ Hazard overhead
✗ Over-reaching up and sideways
✗ No grip on ladder
✗ Sideways-on to work
✗ Foot on handrail
✗ Wearing slippers
✗ Loose tools on ladder
✗ Slippery and damaged steps
✗ Uneven soft ground
✗ Damaged stiles
✗ Non-slip rubber foot missing

Steps at right height ✓
No need to over-reach ✓
Good grip on handrail ✓
Working front-on ✓
Wearing good flat shoes ✓
Clean undamaged steps ✓
Firm level base ✓
Undamaged stiles ✓
Rubber non-slip feet all in position ✓
Meets British or European standards ✓

Figure 4.27 Working with step ladders.

4.9.5 Roof work

Access to and work on roofs is often dangerous and requires adequate precautions to reduce risks to acceptable levels.

There should be a strict procedure for roof work, which often requires the issue of a written safety procedure.

4.9.6 Specialist access equipment

There are many different pieces of specialist access equipment such as pre-fabricated scaffold towers, and mobile elevating work platforms. These should only be used with great care and by properly trained people. This type of equipment can often be hired out and it is tempting to use the equipment without the proper training. Reputable hire companies will check on the level of training and experience that hirers have before letting the equipment out.

4.10 Confined spaces

Entry into confined spaces is potentially very dangerous and a special risk assessment is required. Such areas include tanks, pits, vats, ovens, etc. and perhaps even a cellar. There may be a lack of oxygen or dangerous gases present.

Precautions involve special training, testing the atmosphere, a safe system of work like a permit to work, special tools and rescue equipment. Work is covered by special regulations (Confined Space Regulations) and should be left to specially trained people.

☞ *Safe work in confined spaces* INDG258 available free from the HSE

Appendix 4.1

Manual handling risk assessment

Employee checklist
Task description...

....

Employee ID No

Risk Factors

A. Task characteristics	Yes/No	Risk level			Current controls
		H	M	L	
1. Loads held away from trunk?					
2. Twisting?					
3. Stooping?					
4. Reaching upwards?					
5. Extensive vertical movements?					
6. Long carrying distances?					
7. Strenuous pushing or pulling?					
8. Unpredictable movements of loads?					
9. Repetitive handling operations?					
10. Insufficient periods of rest/recovery?					
11. High work rate imposed?					
B. Load characteristics					
1. Heavy?					
2. Bulky?					
3. Difficult to grasp?					
4. Unstable/unpredictable?					
5. Harmful (sharp/hot)?					
C. Work environment characteristics					
1. Postural constraints?					
2. Floor suitability?					
3. Even surface?					
4. Thermal/humidity suitability?					
5. Lighting suitability?					
D. Individual characteristics					
1. Unusual capability required?					
2. Hazard to those with health problems?					
3. Hazard to pregnant workers?					
4. Special information/training required?					

Any further action needed?   Yes/No
Details:

Chapter 4

Manual handling of loads: assessment checklist

Section A – Preliminary

Job description:	Is an assessment needed?
	(i.e. is there a potential risk of or injury, and are the factors beyond the limits of the guidelines?)
Factors beyond the limits of the guidelines?	
	Yes/No

If 'YES' continue. If 'NO' the assessment need go no further.

Operations covered by this assessment (detailed description):	Diagrams (other information)
Locations:	
Personnel involved;	
Date of assessment	

Section B – See separate sheet for detailed analysis

Section C – Overall assessment of the risk of injury?   Low/Medium/High

Section D – Remedial action to be taken:

Remedial steps that should be taken, in order of priority:

1

2

3

4

5

6

7

8

Date by which action should be taken:

Date for reassessment:

Assessor's name: Signature:

Hazardous substances – Health hazards

5

Summary

- Hazardous substances and how to control them
- Asbestos at work
- Dermatitis
- Drug and alcohol policies
- Legionnaires disease
- Personal protective equipment
- Smokefree policies at work

5.1 Hazardous substances

5.1.1 Introduction

Hazardous substances can be found in all sorts of working environments and unless the right precautions are taken, they can threaten the health of workers and others exposed to them. Many substances can harm people if they enter the body. Exposure can have an immediate effect and repeated exposure can damage the lungs, liver or other organs. Some substances may cause asthma and many can damage the skin.

Chemicals can enter the body by:

- coming into contact with skin or eyes either directly through the skin or into cuts or through clothing;
- breathing in dusts, gases and fumes which quickly enter the blood stream through the lungs; or
- swallowing contaminated food or drink or even worse, drinking a chemical improperly labelled or in a normal drink container.

The 'Control of Substances Hazardous to Health (COSHH) Regulations' provide the legal framework to protect people against health risks from hazardous substances used at work. All businesses have to consider how COSHH applies to their work. Many will be able to comply with the Regulations with little effort; others, whose work creates greater risks, will have more to do.

Hazardous substances can normally be identified by the orange and black warning symbols on the containers, which are classified as falling into three distinct groups as shown in Fig. 5.1.

COSHH covers most substances hazardous to health found in workplaces of all types. The substances covered by COSHH include:

- substances used directly in work activities (e.g. solvents, paints, adhesives, cleaners);
- substances generated during processes or work activities (e.g. dust from sanding, fumes from welding);
- naturally occurring substances (e.g. grain dust).

For the vast majority of commercial chemicals, the presence (or not) of a warning label will indicate whether COSHH is relevant. For example, there is no warning label on ordinary household washing-up liquid, so if it's used at work you do not have to worry about COSHH; but there is a warning label on bleach and so COSHH does apply to its use in the workplace.

Row 1
(a)   (b)   (c)

Row 2
(d)   (e)   (f)

Row 3
(g)

Figure 5.1 (Row 1) Hazardous to Health. (Row 2) Harmful by physico-chemical effects. (Row 3) Dangerous for the environment.

But COSHH does **not** include:

- asbestos and lead, which have specific regulations
- substances which are hazardous only because they are:
  - o radioactive
  - o simple asphyxiants
  - o at high pressure
  - o at extreme temperatures
  - o have explosive or flammable properties (separate Dangerous Substances Regulations cover these)
  - o biological agents if they are not directly connected with work and are not in the employer's control, such as catching flu from a workmate.

5.1.2 In order to comply with COSHH employers must

- assess the level of danger for each hazardous substance by determining the effects of each substance used on the premises. This is called a COSHH Assessment and will include taking precautionary measures;
- decide what precautions are needed in order to prevent anyone from being exposed, for example safety equipment, clothing or ventilation;
- choose the best control measures for each substance, for example, limiting use to a number of people, reducing time of exposure or

using the substance in a controlled area. The principles of good practice set out in the regulations should be followed – see box;

◆ monitor the exposure of workers to hazardous substances. Ensure that all control measures are in place and are being implemented;

◆ train, instruct and inform all relevant employees on how to use hazardous substances, including which cleaning, storage/disposal and emergency procedures to follow.

☞ *COSHH a brief guide to the regulations* INDG136 (Rev2)

☞ *Read the label* INDG186

The principles of good practice are:

1. Design and operate processes and activities to minimise the emission, release and spread of substances hazardous to health.

2. Take into account all relevant routes of exposure – inhalation, skin and ingestion – when developing control measures.

3. Control exposure by measures that are proportionate to the health risk.

4. Choose the most effective and reliable control options that minimise the escape and spread of substances hazardous to health.

5. Where adequate control of exposure cannot be achieved by other means, provide, in combination with other control measures, suitable personal protective equipment (PPE).

6. Check and review regularly all elements of control measures for their continuing effectiveness.

7. Inform and train all employees on the hazards and risks from substances with which they work, and the use of control measures developed to minimise the risks.

8. Ensure that the introduction of measures to control exposure does not increase the overall risk to health and safety.

5.1.3 COSHH assessments

Appendix 5.1 can be used to make the COSHH assessments. The following stages are involved:

(a) List all hazardous substances

◆ Make an *Inventory* of all hazardous substances stored, used or produced by your work. This should list what is stored or produced, in what form, in what quantities and where.

♦ Look for *Hazard Symbols* which classify the substance as Very Toxic, Toxic, Corrosive. Harmful or Irritant. Check up on unmarked or unclearly marked containers which you suspect may contain hazardous contents (Fig. 5.2).

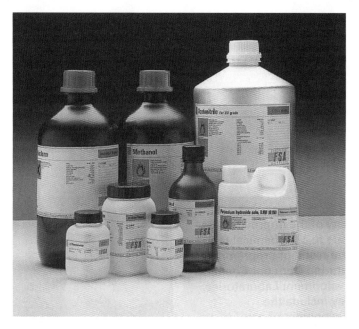

Figure 5.2 Hazardous substances.

♦ Include dusts, fumes or by-products which have Workplace Exposure Limits set in *EH40* published by Health and Safety Executive (HSE) Books.

(b) Find out more about their hazardous nature

♦ Look at the labels on the containers. They must provide basic information on the substances and safety requirements (Fig. 5.3).

♦ Obtain *Hazard Data Sheets* from product suppliers if the labels do not provide sufficient information. These will provide more information on the products hazards, conditions of use and some emergency advice (Fig. 5.4).

(c) How are these substances used?

♦ Examine the work process carefully. Some areas you may need to look at are: delivery and storage; preparation and use; other activities taking place nearby (e.g. cleaning, maintenance, eating, smoking); disposal of residues and containers; unplanned events and emergencies.

Chapter 5

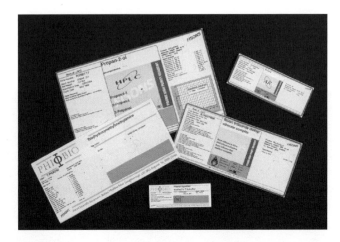

Figure 5.3 Labels.

Figure 5.4 Data sheets.

(d) What are the risks of injury or ill health?

- ◆ From your knowledge of the substance, the work process and foreseeable emergencies decide what the chances are of somebody being injured or suffering from ill health effects and how this could occur. Only significant risks need be considered.

- ◆ Risks could arise from a number of sources, for example: spillages and splashes; mixing incompatible substances; breathing dusts and vapours; absorption through skin contact with the substance; accidental ingestion.

- ◆ Consider who is at risk. Include third parties such as visitors, the public and contractors on your premises in addition to your staff. The length of exposure is also important.

(e) Decide on the precautions needed

◆ The easiest way to remove the risk from a hazardous substance is to stop using it altogether (elimination) or replace it with another less harmful product (substitution).

◆ Thereafter, consider removing those at risk from the substance (enclosure and isolation) or extracting the dusts and fumes from the work area (dilution and local exhaust ventilation).

◆ As an absolute LAST RESORT direct protection of the worker should be considered (PPE).

◆ All these precautions are known as Controls and you must ensure that they are being used properly and monitor and maintain their effectiveness. In some cases, health surveillance must be considered, but this is only in special cases.

(f) Record your assessment

◆ Unless the assessment is simple and you can recall and explain its conclusion at any time, you should put it in writing. The assessment should define clearly who is at risk and why, and what precautions are in place to reduce the risks to a reasonable level.

(g) Instruct and train your staff

◆ Staff need to be told about the *hazardous nature* of the substances they are working with and any risks to which they may be exposed.

◆ Instruction and training must be given on any precautions which must be taken; control measures and their correct use; and what PPE and clothing is needed and how it should be properly used.

◆ Training on Emergency Procedures must also be given to staff.

◆ Remember to keep records on who has been trained and when.

(h) Review your assessment

You need to review the assessment on a regular basis (not more than 5 yearly intervals). The assessment must also be reviewed if there is a change in product or process. The assessment should state when it is to be reviewed next.

5.2 Asbestos

5.2.1 Introduction

Since about 1950 asbestos has been the main cause of occupational ill health and it is still the greatest single work-related cause of death.

About 4000 people die every year from asbestos related cancers resulting from past exposure. This figure is expected to rise over the next ten years and then decline.

The cost in human suffering is enormous. For the people involved there is immense pain and suffering and their relatives, friends and colleagues suffer with them.

5.2.2 What is it?

Asbestos is the name used for a range of natural minerals of which there are three main types:

- ◆ blue (crocidolite);
- ◆ brown (amosite);
- ◆ white (chrysotile).

You cannot identify the type of asbestos just by its colour.

Asbestos has been used in a very large number of products, especially in buildings. Some of the products have one type of asbestos in them while others have mixtures of two or more (Fig. 5.5).

Figure 5.5 Asbestos removal sign.

There is danger in all types of asbestos.

5.2.3 What makes asbestos dangerous?

Asbestos is made up of thin fibres which can break down into much smaller, finer fibres. The finest ones are too small to see with the naked eye but they can be breathed in.

ALL types of asbestos fibres are potentially fatal if breathed in but this can only happen if they are made airborne first.

When these fine fibres are breathed in they get stuck in the lungs and damage them. This causes scars. The scars stop the lungs working properly (asbestosis), or they can cause cancer. The main types of cancer caused by asbestos are cancer of the lung and cancer of the lining of the lung (mesothelioma).

It can take from 15 to 60 years for these diseases to develop and there is no cure for any of them.

5.2.4 Where am I likely to find asbestos?

Buildings built or refurbished between 1950 and 1980 are the places where you are most likely to find asbestos. Thousands of tonnes of asbestos products were used in buildings around that time and a lot of it is still there. But it's difficult to identify the products from their appearance.

Because it was such an effective material, it was very widely used, in the following ways:

- ◆ Fire insulation, known as 'limpet', was sprayed on structural steel beams and girders;
- ◆ It was used to lag, pipework, boilers, calorifiers, heat exchangers, insulating jackets for cold water tanks and around ducts;
- ◆ Asbestos insulation board (AIB), was used for ceiling tiles, partition walls, soffits, service duct covers, fire breaks, heater cupboards, door panels, lift shaft linings, fire surrounds;
- ◆ Asbestos cement (AC), was used for roof sheeting, wall cladding, walls and ceilings, bath panels, boiler and incinerator flues, fire surrounds, gutters, rainwater pipes, water tanks;
- ◆ It was loosely packed between floors and in partition walls;
- ◆ There were many other products, for example, floor tiles, mastics, sealants, textured decorative coatings, rope seals, gaskets, millboards, paper products, fire doors, cloth (e.g. fire blankets), and bituminous products (roofing felt) (Fig. 5.6).

5.2.5 Am I at risk of being exposed to asbestos fibres?

If you disturb asbestos-containing materials, for example, by working on them or near them, you are likely to be exposed.

Research suggests that the groups most at risk are people who do building maintenance and refurbishment work. If you work as a demolition contractor, electrician, construction contractor, heating and ventilation engineer, then you are likely to be at risk from asbestos fibres.

Chapter 5

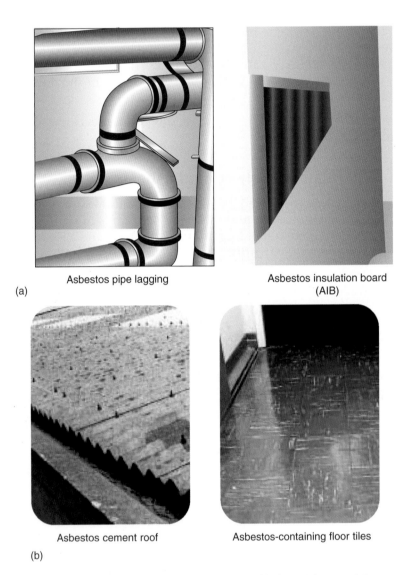

Asbestos pipe lagging

Asbestos insulation board
(AIB)

(a)

Asbestos cement roof

Asbestos-containing floor tiles

(b)

Figure 5.6 (a) High risk materials and (b) normally low risk materials.

5.2.6 New asbestos regulations 2006

Most of the duties under the Control of Asbestos Regulations 2006 are the same as they were before, but there are some important changes:

◆ The new control limit is lower. This is the limit over which no one must go. It is 0.1 fibres per millilitre of air measured over 4 hours.

- Generally, you will not need to have work done by a licensed contractor if it is with textured coatings, but it will still need to be done safely by trained, competent people working to certain standards.
- If the work would otherwise require a licence, employers can no longer carry out work in their own premises with their own workers without a licence;
- The part of the Regulations that deals with training is clearer than before. It says that suitable training is required for anyone who is, or may be, exposed to asbestos.

5.2.7 When do I need a licence?

The most dangerous asbestos-containing materials give off high fibre levels when disturbed. For work with these you will need a licence from the HSE. In fact work with most asbestos-containing materials requires a licence.

Virtually all work with loose packing, sprayed insulation, lagging and AIB needs a licence. If you are doing a job which, in total, takes one person no more than 1 hour, or more people no more than 2 hours in any 7-day period, then you will not need a licence. But this would only apply to very minor pieces of work.

If you have made a risk assessment and confirmed that the exposure (without a respirator) will not go above 0.6 fibres per millilitre in any 10-minute period or go over the control limit and the work involves certain materials, then a licence is not needed. So, for work involving textured coatings, AC, and other materials where the fibres are firmly held in the material, for instance vinyl floor tiles and bituminous products such as roofing felt, you will generally not need a licence.

5.2.8 What do I need to do about the regulations?

As an employer, this section tells you what questions you should be asking yourself and tells you a little bit more about the regulations.

The Regulations apply to all work with asbestos materials carried out by employers, the self-employed and employees. They apply to all work with asbestos whether it requires a licence or not.

If you have control of a building you have a duty to manage any asbestos in it, so you must take reasonable steps to discover whether there are materials containing asbestos in the premises. If there are, you need to know how much, where they are and what condition they are in. This may or may not involve a survey.

What would a survey involve?

There are three types of survey:

- ◆ Type 1 is known as presumptive. In this type of survey the materials assumed to contain asbestos are located and note is taken of their condition. No sampling is done.
- ◆ Type 2 is known as sampling. The survey is the same as type 1 but samples are taken and analyzed to confirm whether asbestos is present.
- ◆ Type 3 is known as full access. Full access to all parts of the building will be needed using destructive inspection if necessary. This type is usually used just before demolition or major refurbishment.

You must record the results of all types of survey and the information must be given to anyone who may work on, or disturb, these materials.

Before you start any work which could disturb asbestos, or may expose employees to it, you will need to make a risk assessment. A competent person should do the assessment. The person who controls the premises needs to manage the risk and will have to make sure that they know whether there is, or may be, any asbestos in the building. The assessment must include where the asbestos is, or is thought to be, and what condition it is in. You should always assume that asbestos could be present until a full survey is done.

Do not carry out demolition, maintenance or any other work which exposes, or may expose, your employees to asbestos in any premises unless you know:

- ◆ whether asbestos is, or may be, present;
- ◆ what type of asbestos it is;
- ◆ what material it is in; and
- ◆ what condition it is in; or
- ◆ if there is any doubt about whether asbestos is present, you must assume that it is present and that it is not only white asbestos.

This training needs to be given at regular intervals. It should be in proportion to the nature and degree of exposure. It should contain the appropriate level of detail and be suitable to the job. Use written, oral and practical methods to get the information across.

All asbestos-containing materials should be clearly marked, even if in good condition. There is only a risk if asbestos fibres are made airborne. Usually this happens if asbestos materials are damaged or disturbed.

Material which may contain asbestos and has been damaged needs special attention. It's important not to create more of a risk to people by, for example, causing a panic or leaving something in an unsafe condition, but if you see material which you think contains asbestos, and it has been damaged so that you believe that there is a serious risk of exposure to asbestos fibres, you should ask everyone to leave the area. Minor damage to some asbestos materials does not always mean that there is a serious risk or that immediate evacuation of the area is warranted. If the materials are securely bound in a matrix, textured coatings or AC for instance, and damage is only slight, it is less of a risk, but damaged edges should be coated immediately, and repaired as soon as possible (Fig. 5.7).

Chapter 5

Figure 5.7 Asbestos removed and repainting.

5.2.9 How can I get more information about this?

Call HSE's Infoline for confidential advice and information (you do not have to give your name) on 0845 345 0055.

See the HSE website: www.hse.gov.uk/asbestos.

Asbestos essentials task manual: Task guidance sheets for the building maintenance and allied trades HSG210 ISBN 0 7176 1887 0 available from HSE Books.

5.3 Dermatitis

5.3.1 Introduction

This debilitating and unsightly condition of the skin causes irritation and frequently pain as well. It can be seriously disabling and even bring an end to a person's employment prospects. Early signs are redness, flaking, itching and cracking of the skin, particularly in the webs between the fingers. Because they are the areas most likely to touch a substance, the hands and forearms are most often affected; but it is possible to get dermatitis on the neck, face and chest too and if the skin cracks and bleeds, dermatitis can spread all over the body. However it is important to understand that it is not infectious (Fig. 5.8).

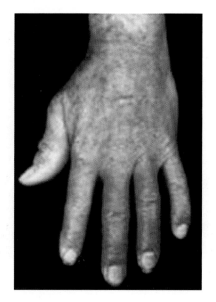

Figure 5.8 Dermatitis on hand.

Occupational dermatitis costs UK employers up to 85 million a year and there is also a high cost in human suffering. This includes the loss of a social life, inability to use the hands because of the intense pain and no longer being able to pursue a chosen career. The business sectors with the highest risks are hairdressing and beauty, the construction, engineering, printing, chemicals, health care and residential care, catering, cleaning, horticulture, gardening and floristry industries (Fig. 5.9).

5.3.2 What you must do – make an assessment of the risk

If you employ people you have a legal duty to prevent your employees from developing dermatitis because of the substances they are using at work, so you will need to assess the risk.

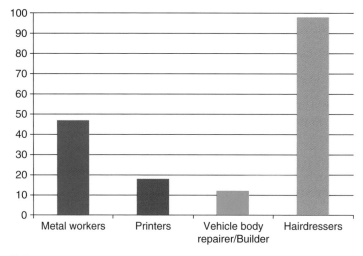

Figure 5.9 Incidence of contact dermatitis among industry sectors. Incidence rate per 100,000. *Source*: HSE.

The first thing to do is to ask yourself four questions:

1. *Firstly, does my business have a problem?*

   If you or anyone working with you has a skin problem, it is possible that one or more of the substances that you are using is causing it.

2. *Secondly, what are we using that might cause a problem?*

   Take a look at the labels and safety data sheets for the substances that you are using. If you haven't got them, ask the supplier (they are obliged to provide them) or look the information up on the internet. There may be a warning on the label but not every substance carries a warning. Things like shampoo, some metal cleaners and other cleaning materials can cause dermatitis if they are used every day over a long period, but they may not be labelled as a risk.

3. *Thirdly, is there something safer we could use?*

   Find out from the HSE or a trade association whether you could use a safe alternative.

4. *Fourthly, is there a safer way to do the job?*

   Stopping skin contact by using gloves or other protective clothing (also see Section 5.1), changing to an automated process or enclosing the process; or in the case of fumes and dust, providing exhaust ventilation could be the solution

Some substances can either irritate the skin (e.g. acids or alkalis, detergents or degreasers (solvents), or sensitize the skin (e.g. some ingredients found in perfumes or soaps, hair dyes, components of some synthetic rubber gloves).

People can develop dermatitis after a single exposure to an irritating substance, or after repeated exposure over a long period of time, since over time the skin becomes less resilient.

The concentration of a substance is important and so is the length of time people are in contact with it. Substances that are used very frequently are more likely to cause problems.

You will need regular checks to pick up on early symptoms. Like most things, the earlier you diagnose the condition, the more likely it is that you will be able to alleviate it .

5.3.3 Controlling the risk of dermatitis

Although in some jobs there is nothing you can do to stop contact with substances that cause dermatitis, there is a lot you can do to control the risk. Of course if you can prevent the problem, and avoid using irritating and sensitising substances altogether, that is a much better solution, but if that is not possible there are several ways of providing protection (Fig. 5.10):

◆ People should be told about the risks of dermatitis if they are using substances that are likely to cause it. They should be trained to use the substances properly.

◆ The right type of protective clothing and gloves should be provided. There are different types for different substances and jobs. It's important to get this right and you can ask the supplier about it. Where liquids, fumes and dust might get on to the face or neck, face masks and shields may be necessary.

◆ Unless you are using disposable clothing and gloves, you will need to get them cleaned and replace them regularly.

◆ You will need to provide adequate washing facilities – keeping the skin clean is important.

◆ Encourage people to use moisturising cream as it helps to replace the loss of natural oils.

◆ If you are diluting substances, make sure they are diluted to the right strength. If they are too strong then they are more likely to cause dermatitis.

◆ Ideally, avoid using strongly irritating or sensitising substances altogether by changing the process or using a less harmful substitute.

◆ Limit the number of people who are exposed to the substance; reduce the time and frequency that they are exposed.

◆ Keep the workplace and any machinery or tools clean.

Figure 5.10 Bad hand day. HSE website.

5.3.4 Training and supervision

- ◆ If people are using substances likely to cause dermatitis, they need to be aware of the early signs of sensitisation and irritation.
- ◆ Make sure that they know all the precautions that they should take and what to do if they think they have a problem.

◆ Make sure that people are using the substances properly (e.g. diluting to the right strength) and that they are wearing or using any protective clothing or equipment that has been provided.

5.3.5 Setting up a system for early signs of skin problems

You will need to set up a system of regular checking of exposed skin to identify the early signs of dermatitis. This is an important part of your overall control of the risk. Early detection along with the right treatment will halt and may reverse the process.

To begin with you will probably need an occupational health nurse or a doctor to advise you as to how often you need to make checks. They will also need to train someone to carry out the checks. Once trained, this person will only be able to look for abnormal signs, since they will not be qualified to make a judgement about the cause of a skin problem. Anyone who appears to have a skin problem must be referred to their GP, or, if you have one, the works occupational nurse or doctor.

5.3.6 Report cases to your enforcing authority

Cases of dermatitis (diagnosed by a registered medical practitioner) must be reported to your enforcing authority on form F2508a where they are caused by the use of certain specified substances in the workplace, including any known irritant or sensitising agent.

Other help available:

'Rash Decisions' HSE video on work related dermatitis – its causes, effects and prevention

Leaflet INDG 233 'Preventing dermatitis at work'. Also Hairdressing Section – Bad Hand day campaign: HSE website.

5.4 Drug and alcohol policy at work

A drug is any substance, which when it enters the Central Nervous System, alters mind or body. This **does** include alcohol.

5.4.1 What should a drug and alcohol policy do?

◆ Alert staff to the problems associated with the misuse of drugs

8

- Offer encouragement and assistance to all employees who feel they may have a drug problem to seek voluntary help at an early stage.
- Offer assistance to an employee with a drug related problem that comes to light through observation or through the normal disciplinary procedure, for example, through poor work performance, absenteeism or misconduct.

5.4.2 Training

The success of a company's drug abuse policy depends on how well and how seriously that policy is communicated to employees. Even the best policy can be completely ineffective if people don't know it exists. Copies of the policy should be distributed to everyone with an accompanying letter from a senior member of management. Copies could also be displayed on public notice boards within the organisation to serve as a reminder.

Supervisors have the most contact with staff and they should have the knowledge to enable them to spot problems. They need specific knowledge of all kinds of drugs and/or alcohol and their possible effects so you may need to provide information and training to enable them to implement the policy.

5.4.3 Decide if you have a problem

Indicators of drug and/or alcohol impairment include:

- Performance problems
- Physical appearance (side effects)
- Lack of co-ordination
- Inappropriate mood

Once a problem is suspected the supervisor needs to keep records of any drug and/or alcohol related incidents. This will help you to challenge the person who is abusing drugs and/or alcohol (Fig. 5.11).

5.4.4 Address the problem

The second step is to address the problem. Be calm and assertive. Four suggestions for the interview are:

- Begin by expressing concern over the change in the employee's work performance.
- Use the records you have kept to back up your position.

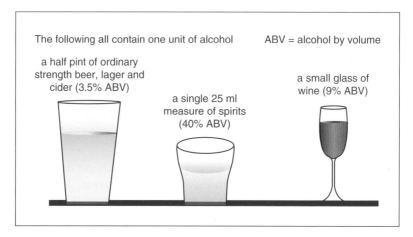

The following all contain one unit of alcohol   ABV = alcohol by volume

a half pint of ordinary
strength beer, lager and
cider (3.5% ABV)

a single 25 ml
measure of spirits
(40% ABV)

a small glass of
wine (9% ABV)

Figure 5.11 Alcohol consumption: It takes a healthy liver about 1 hour to break down and remove 1 unit of alcohol. A unit is equivalent to 8 gm or 10 ml (1 cl) of pure alcohol. If someone drinks 2 pints of ordinary strength beer at lunchtime or half a bottle of wine (i.e. 4 units), they will still have alcohol in their bloodstream 3 hours later. Similarly, if someone drinks heavily in the evening they may still be over the legal drink drive limit the following morning. Black coffee, cold showers and fresh air won't sober someone up. Only time can remove alcohol from the bloodstream. *Source*: HSE.

◆ Clearly express expectations for change in work behaviour and offer possible suggestions.
◆ Offer help and support.

If an employee refuses treatment, their work performance should be monitored for a specified period. If it remains unsatisfactory the employee should be interviewed again and if necessary disciplinary procedures should be invoked.

If an employee accepts help he or she should be asked to sign an agreement setting out obligations on both sides. You may need outside help for this. There are drug and alcohol advisory centres in most areas whose staff will be able to point you in the right direction. Your local health centre is also likely to be a good source of help and information.

If the problem has arisen because of the nature of the work then, where possible, the person should be redeployed.

The employee should be offered confidentiality during treatment. No personnel record should be kept of the employee undergoing treatment, although, if you have a medical department they will keep medical records.

If a drug and/or alcohol problem recurs either during treatment or later, then each case should be assessed on its merits. It is possible that further treatment could be offered but this is increasingly unlikely to be effective.

5.5 Legionnaires' disease

5.5.1 The problem

Legionella bacteria are widespread in natural sources of water including rivers, streams and ponds and may even be found in soil. They are also found in many recirculating hot and cold water systems and in some cooling towers attached to air conditioning systems. Outbreaks of legionnaires' disease have occurred in or near large complexes such as hotels, hospitals, offices and factories. There is no evidence that water systems in domestic homes present a risk.

The symptoms of legionnaires' disease are similar to the symptoms of flu:

◆ high temperature, feverishness and chills;
◆ cough;
◆ muscle pains;
◆ headache;

leading on to pneumonia, very occasionally diarrhoea and signs of mental confusion.

The illness is treated with an antibiotic called erythromycin or a similar antibiotic.

Those most at risk include smokers, alcoholics and people suffering from cancer, diabetes, chronic respiratory or kidney disease. But healthy people can also be infected. Most reported cases have been in people aged between 40 and 70; men are more likely to be affected than women.

Fortunately the chain of events necessary for susceptible people to be infected does not happen very often.

5.5.2 Reducing the risks

Hot and cold water systems should not have sections where water stands for long periods undisturbed. Cisterns should be covered and periodically inspected, cleaned and disinfected. Water temperatures between 20°C and 45°C should be avoided by insulating cold pipes/tanks in warm spaces and storing hot water at 60°C and circulating at 50°C. Where there is risk of scalding thermostatically controlled taps should be used.

Cooling towers should be well designed, maintained and operated. Specialist assistance will be needed to give appropriate advice. Where practicable they should be replaced with dry cooling systems.

5.6 Personal protective equipment

5.6.1 General requirements

Personal Protective Equipment at Work Regulations 1992 (PPE): The Regulations cover all equipment (including clothing to protect against the weather), which is intended to be worn or held by a person at work and which protects them against one or more risks to their health and safety. The effectiveness of PPE can easily be defeated. For example, it may not be worn properly or a second unprotected person may be close by, so it should always be considered as a last resort and used only when other precautions will not adequately reduce the risk.

However waterproof, weatherproof or insulated clothing is covered only if its use is necessary to protect against adverse climatic conditions. Ordinary working clothes and uniforms, which do not specifically protect against risks to health and safety, and protective equipment worn in sports competitions are not covered (Fig. 5.12).

Figure 5.12 Tree surgeon with eye protection, ear muffs, helmet and harness.

Where it is necessary to protect people against health and safety risks, all necessary PPE must be provided free of charge to employees. The employer also has a duty to maintain and clean the equipment and will need to supply replacements when needed.

Where there is overlap in the duties in these Regulations and those covering lead, ionising radiations, asbestos, hazardous substances (COSHH), noise, and construction head protection then the specific requirements in those regulations should be followed.

When appropriate protective clothing or equipment needs to be provided the following precautions should be taken:

- Ensure that equipment/clothing is suitable and appropriate for the hazard that is being protected against. For example, eye protection designed to protect against a chemical splash will not provide protection against particles thrown off from the use of an angle grinder.
- Ensure that it prevents or properly controls the risk without increasing the overall level of risk.
- Ensure the equipment/clothing is of good quality made to a recognised standard, for example, BS/EN/ISO/DIN as appropriate, look for the CE marking.
- Ensure the equipment/clothing suits the wearer in size, weight and fit.
- Ensure that the state of health of the wearer has been considered.
- Consider the needs of the task and the extra strain it may put on the wearer.
- Consider the compatibility of different PPE for example goggles and the fit of a face mask.

You must instruct and train employees in the use of the equipment/clothing. Tell them why it is needed, when to use it and what its limitations are. Ensure that all equipment/clothing is properly looked after and stored when not in use. It must be kept clean and in good repair. Ensure that staff actually wear the equipment/clothing provided. Both employers and employees have legal responsibilities to ensure that it is worn. Employees must report defects in their equipment/clothing to their employers immediately so that appropriate steps can be taken.

5.6.2 Examples of PPE

(a) EYES

Hazards: chemical or metal splash; dust; projectiles; gas and vapour; radiation.

Choices: spectacles; goggles; face-screens.

(b) HEAD AND NECK

Hazards: impact from falling or flying objects; risk of head bumping; hair entanglement.

Choices: helmets; bump caps; hats; caps; sou'westers and cape hoods; skull-caps.

(c) BREATHING

Hazards: dust; vapour; gas; oxygen deficient atmospheres.

Choices: disposable filtering face piece or respirator; half/full face respirators; air-fed helmets; breathing apparatus.

(d) PROTECTING THE BODY

Hazards: temperature extremes; adverse weather; chemical or metal splash; spray from pressure leaks or spray guns; impact or penetration; contaminated dust; excessive wear or entanglement of own clothing.

Choices: conventional or disposable overalls; boiler suits; donkey jackets; specialist protective clothing, for example, chain-mail aprons; high visibility clothing; waterproof jacket.

(e) HANDS AND ARMS

Hazards: abrasion; temperature extremes; cuts and punctures; impact; chemicals; electric shock; skin infection, disease or contamination; vibration.

Choices: leather gloves; waterproof gloves; gauntlets; mitts; armlets.

(f) FEET AND LEGS

Hazards: wet; electrostatic build-up; slipping; cuts and punctures; falling objects; metal and chemical splash; abrasion.

Choices: safety boots and shoes with steel toe caps (and steel mid sole); gaiters; leggings; spats

5.7 Smokefree workplaces

5.7.1 Introduction

Secondhand tobacco smoke is a major cause of heart disease and lung cancer amongst non-smokers who work with people who smoke. It is estimated that around 700 workers a year die as a direct result of second-hand tobacco smoke in their workplace.

> Secondhand smoke contains over 4,000 chemicals. Over 50 are known to cause cancer.
>
> Around 85% of secondhand smoke is invisible and odourless.

Secondhand smoke has been responsible for many thousands of episodes of illness. For example, Asthma UK reported that it was the second most common asthma trigger in the workplace. Eighty-two per cent of people with asthma reported that other people's smoke worsened their asthma and in the past 1 in 5 people with asthma felt excluded from parts of their workplace where people smoke.

Before July 2007, around a quarter of workers smoked, although not necessarily in the workplace, where there has been a steady move towards smoking restrictions over the past 20 years. As a result most workers already worked in a 'smokefree' environment. However, around 2 million people in Great Britain were still working in workplaces where smoking was allowed throughout, and another 10 million in places where smoking was allowed somewhere on the premises. Nicotine is extremely addictive and many smokers find adjusting to smoking restrictions difficult.

From July 2007 legal restrictions on smoking in workplaces and public places have been in force throughout the UK. Similar legislation had already been implemented in Wales, Scotland and Northern Ireland. The legislation effectively bans smoking in all enclosed workplaces and public places, with some exemptions. The following comes from the Department of Health website to ensure that it is correct for this radical new piece of legislation.

5.7.2 A quick guide to the new smokefree law is as follows:

◆ England became smokefree on Sunday, 1 July 2007. The new law was introduced to protect employees and the public from the harmful effects of secondhand smoke.

◆ From 1 July 2007 it is against the law to smoke in virtually all enclosed public places, workplaces and public and work vehicles. There will be very few exemptions from the law.

◆ Indoor smoking rooms in virtually all public places and workplaces will no longer be allowed.

Chapter 5

- Managers of smokefree premises and vehicles have legal responsibilities to prevent people from smoking.
- The new law requires no-smoking signs to be displayed in all smokefree premises and vehicles.
- The new law applies to anything that can be smoked. This includes cigarettes, pipes (including water pipes such as shisha and hookah pipes), cigars and herbal cigarettes.

Failure to comply with the new law is a criminal offence.

Penalties and fines for smokefree offences are set out below (there are some discounted amounts for quick payment):

- **Smoking in smokefree premises or work vehicles**: A fixed penalty notice of £50 imposed on the person smoking. Or a maximum fine of £200 if prosecuted and convicted by a court.
- **Failure to display no-smoking signs:** A fixed penalty notice of £200 imposed on whoever manages or occupies the smokefree premises or vehicle. Or a maximum fine of £1000 if prosecuted and convicted by a court.
- **Failing to prevent smoking in a smokefree place:** A maximum fine of £2500 imposed on whoever manages or controls the smokefree premises or vehicle if prosecuted and convicted by a court.

Local councils will be responsible for enforcing the new law. They will offer information and support to help businesses meet their legal obligations under the new law.

You can find out more information on the new law on the Smokefree England website at **smokefreeengland.co.uk**. You can also contact your local council for information.

5.7.3 Getting ready to go smokefree

If you manage or are in charge of any premises or vehicles that the new law applies to you will have a legal responsibility to ensure they become and remain smokefree. Comply with the new law which came into effect on Sunday, 1 July, you'll need to make sure that:

- you have all the required no-smoking signs in place (see Appendix 5.3 page 173);
- your staff, customers, members or visitors are aware that your premises and work vehicles are legally required to be smokefree; and
- you have removed any existing indoor smoking rooms.

5.7.4 Keeping your premises and vehicles smokefree

From July 2007, it has been the legal responsibility of anyone who controls or manages smokefree premises and vehicles to prevent people from smoking in them. You will have to demonstrate that you have taken reasonable steps to meet the requirements of the new law. These might include:

(a) removing ashtrays

(b) introducing a smokefree policy (see below)

(c) training staff to understand the new law and what their responsibilities are.

5.7.5 Introducing a smokefree policy for your business

You may wish to introduce a smokefree policy for your workplace. This will help ensure your employees are aware of the new smokefree law and that they now work in a smokefree environment. It will also advise them on what they need to do to comply with the new law.

Your smokefree policy should be developed in consultation with employees and their representatives.

The smokefree policy can be a verbal understanding between you and your employees, incorporated into your existing corporate or health and safety policies, or you may wish to create a separate written policy.

We have enclosed a typical smokefree policy which is suggested by the Department of Health.

5.7.6 Is there support available for anyone who wants to stop smoking?

Around 70% of smokers say that they want to stop smoking, and the new smokefree law could provide extra motivation to do so. If you'd like to help your staff and customers become non-smokers, there is excellent free support available from the NHS. This includes:

Local NHS stop smoking services – To find your local service, call the NHS Smoking Helpline free on 0800 169 0 169, visit **gosmokefree.co.uk**, text 'GIVE UP' and your full postcode to 88088 or ask at your local GP practice, pharmacy or hospital.

NHS smoking helpline – Individuals can speak to a specialist adviser by calling 0800 169 0 169 (lines are open daily from 7 a.m. to 11 p.m.).

gosmokefree.co.uk – An online resource for all the advice, information and support needed to stop and stay stopped.

Together – This support programme is free to join, and is designed to help individuals stop smoking using both medical research and insights from exsmokers. You can choose to receive emails, text messages, mailing packs and phone calls. Call the NHS Smoking Helpline on 0800 169 0 169 or visit **gosmokefree.co.uk** for details.

Employers can speak to their local NHS Stop Smoking Service about support for their employees during or outside working hours.

5.7.7 Which places must be smokefree?

Smokefree public places and workplaces

The new smokefree law applies to virtually all 'enclosed' and 'substantially enclosed' public places and workplaces. This includes both permanent structures and temporary ones such as tents and marquees. This also means that indoor smoking rooms in public places and workplaces are no longer allowed.

Premises are considered **'enclosed'** if they have a ceiling or roof and (except for doors, windows or passageways) are wholly enclosed either on a permanent or temporary basis.

Premises are considered **'substantially enclosed'** if they have a ceiling or roof, but have an opening in the walls, which is less than half the total area of the walls. The area of the opening does not include doors, windows or any other fittings that can be opened or shut (Fig. 5.13).

There is no requirement for outdoor smoking shelters to be provided for employees or members of the public.

If you do decide to build a shelter, we suggest you discuss any plans you may have with your local council, as there may be a range of issues you need to consider. These might include planning permission, licensing, building control, noise and litter.

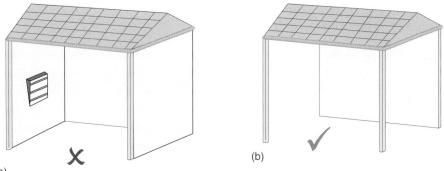

Figure 5.13 (a) Example of substantially enclosed premises. (b) Example of non-substantially enclosed premises.

5.7.8 Smokefree vehicles

The new law requires vehicles to be smokefree at all times if they are used:

(a) to transport members of the public.

(b) in the course of paid or voluntary work by more than one person – regardless of whether they are in the vehicle at the same time.

Smokefree vehicles need to display a no-smoking sign in each compartment of the vehicle in which people can be carried. This must show the international no-smoking symbol no smaller than 70 mm in diameter.

When carrying persons, smokefree vehicles with a roof that can be stowed or removed are not required to be smokefree when the roof is completely removed or stowed.

Vehicles that are used primarily for private purposes are not required to be smokefree.

It is the legal responsibility of anyone who drives, manages or is responsible for order and safety on a vehicle to prevent people from smoking.

5.7.9 Private dwellings

In general, the new law does not cover private dwellings. However, any enclosed or substantially enclosed part of a premises shared with other

Chapter 5

premises, such as a communal stairwell or lift in a block of flats, are required to be smokefree if:

(a) it is open to the public.

(b) it is used as a place of work, for example, by a cleaner, postman or security guard.

The law does not require self-contained residential accommodation for temporary or holiday use (e.g. holiday cottages or caravans) to be smokefree. The owners, however, may choose to make the accommodation smokefree.

Anyone who visits private dwellings as part of their work, for example, delivering goods, or providing services such as plumbing, building or hairdressing, can download further guidance at **smokefreeengland. co.uk/resources**

5.7.10 Working from home

Any part of a private dwelling used **solely** for work purposes is required to be smokefree if:

(a) it is used by more than one person who does not live at the dwelling

(b) members of the public attend to deliver or to receive goods and/or services.

5.7.11 What are the required signs for smokefree premises?

No-smoking signs need to be displayed in a prominent position at every entrance to smokefree premises. Signs must meet the following minimum requirements:

◆ be a minimum of A5 in area (210 mm × 148 mm)

◆ display the international no-smoking symbol at least 70 mm in diameter

◆ carry the following words in characters that can be easily read: '**No smoking. It is against the law to smoke in these premises**'.

You are also free to personalise your signs by changing the words 'these premises' to refer to the name or type of premises – such as 'this gym', 'this salon' or 'this restaurant'.

A smaller sign consisting of the international no-smoking symbol at least 70 mm in diameter may instead be displayed at entrances to smokefree premises that:

- are only used by members of staff – providing the premises displays at least one A5 area sign;
- are located within larger smokefree premises, such as a shop within an indoor shopping centre.

5.7.12 What is the required signage for smokefree vehicles?

Smokefree vehicles need to display a no-smoking sign in each compartment of the vehicle in which people can be carried. This must show the international no-smoking symbol at least 70 mm in diameter.

5.7.13 Where can I get the signs?

We have included a no-smoking sign that meets the requirements of the new law in Appendix 5.3. Additional signs can be downloaded and printed or ordered from **smokefreeengland.co.uk/resources**. Signs can also be ordered from the Smokefree England information line on 0800 169 169 7.

Alternatively, you are allowed to design and print your own no-smoking signs as long as they meet the minimum requirements. The international no-smoking symbol consists solely of a graphic representation of a single burning cigarette enclosed in a red circle of at least 70 mm in diameter with a red bar across it.

Chapter 5

Chapter 5

Appendix 5.1 COSHH assessment forms

COSHH 1 – DETAILS OF SUBSTANCES USED OR STORED

Name of Manager:..

Name of Department/Area:...

SUBSTANCE DETAILS

1. Information from the label

  Trade name:...

  Manufacturer's name:..

  Names of any chemical constituents listed: ....
  ....

  Hazard marking – whether corrosive, irritant, harmful, toxic, very toxic
  ....

  RISKS noted on label (e.g. harmful to eyes) ....
  ....

  PRECAUTIONS noted on label (e.g. use in well-ventilated area) ......
  ....
  ....
  ....

2. Have you got a Health and Safety Data Sheet for this product?
  YES/NO

DETAILS OF USE

3. What it is used for? ....
  ....

4. By whom? ....

5. How often? ....

6. Where? ....

7. What CONTROL measures (precautions) are used? (e.g. local ventilation, goggles, respirator, protective gloves, etc.) ....
  ....
  ....
  ....

8. Is it ABSOLUTELY ESSENTIAL to keep/use this substance? YES/NO

9. Can it be DISPOSED OF NOW?   YES/NO

COSHH 2 – ASSESSMENT OF A SUBSTANCE

1. Name of substance...
2. The process or description of job where the substance is used ....
   ....
3. Location of the process where substance is used ....
4. Health and safety information on substance:
   (a) Hazards to health: ....
   ....
   ....

   (b) Precautions required: ....
   ....
   ....
   ....
   ....
5. Number of persons exposed: ....
6. Frequency and duration of exposure: ....
7. Control measures that are in use: ....
   ....
   ....
   ....
   ....
8. The assessment, an evaluation of the risks to health: ....
   ....
   ....
   ....
   ....
9. Details of steps to be taken to reduce the exposure: ....
   ....
   ....
   ....
   ....
10. Action to be taken by (name): .... (Date):................
11. Date of next assessment/review: ....
12. Name and position of person making this assessment: ....
13. Date of assessment: ....

Appendix 5.2
Smokefree policy

Purpose

This policy has been developed to protect all employees, service users, customers and visitors from exposure to secondhand smoke and to assist compliance with the Health Act 2006.

Exposure to secondhand smoke increases the risk of lung cancer, heart disease and other serious illnesses. Ventilation or separating smokers and non-smokers within the same airspace does not completely stop potentially dangerous exposure.

Policy

It is the policy of Insert name of company that all our workplaces are smokefree, and all employees have a right to work in a smokefree environment. The policy shall come into effect on Sunday, 1 July 2007. Smoking is prohibited in all enclosed and substantially enclosed premises in the workplace. This includes company vehicles. This policy applies to all employees, consultants, contractors, customers or members and visitors.

Implementation

Overall responsibility for policy implementation and review rests with Insert name of manager. However, all staff are obliged to adhere to, and support the implementation of the policy. The person named above shall inform all existing employees, consultants and contractors of the policy and their role in the implementation and monitoring of the policy. They will also give all new personnel a copy of the policy on recruitment/induction.

Appropriate 'no-smoking' signs will be clearly displayed at the entrances to and within the premises, and in all smokefree vehicles.

Noncompliance

Local disciplinary procedures will be followed if a member of staff does not comply with this policy. Those who do not comply with the smokefree law may also be liable to a fixed penalty fine and possible criminal prosecution.

Help to stop smoking

The NHS offers a range of free services to help smokers give up. Visit gosmokefree. co.uk or call the NHS Smoking Helpline on 0800 169 0 169 for details. Alternatively you can text 'GIVE UP' and your full postcode to 88088 to find your local NHS Stop Smoking Service.

Signed ....

Date ....

On behalf of the Company Insert name of company

SMOKEFREE

Appendix 5.3
Smokefree sign

NO SMOKING.

It is against the law to
smoke in these premises

Physical and psychological health hazards

6

Summary

- **Display screen equipment and workstations**
- **Musculoskeletal problems**
- **Noise problems at work**
- **Stress**
- **Vibration at work**
- **Violence and bullying**

6.1 Display screen equipment and computer workstations

6.1.1 Health and Safety (display screen equipment) Regulations 1992

A display screen can be a cathode ray tube like an old-style television, or a flat panel display. The expressions VDU, VDT, monitor and Display Screen Equipment (DSE) all mean the same thing – a display screen. This is usually part of a computer.

The DSE Regulations apply to the workstation, job and work environment and to the VDU, keyboard and related equipment.

Most health and safety problems with VDU's are caused by the way the VDU is used, and you can avoid them by designing the job and the workplace carefully, and making sure that people use their VDU and workstation properly.

People sometimes suffer aches and pains in their hands, wrists, arms, shoulders, neck or back. This has commonly been called repetitive strain injury (RSI) but it is more accurate to describe the whole group of problems as 'upper limb disorders'.

Sometimes these disorders have a physical cause such as working for long periods at a VDU without interruption; but quite often they are caused by stress, a result of pressure to meet deadlines or an increase in the pace of work.

There are several solutions to this. For example, people need to take frequent short breaks from a VDU and they need to be physically comfortable when they are working. Systems and tasks need to be designed to match the needs and abilities of the people working with them, and training needs to be provided.

Using a mouse can cause aches and pains, so people need to adopt a good posture and technique, limiting the time they spend using it and taking breaks. The mouse should be used close to the body with the wrist straight and the arm relaxed and supported. There are several different designs of mouse on the market and it's a good idea to try them out to see which one suits you best.

If you spend a lot of time using a laptop, make sure it is on a firm surface and at the right height to use the keyboard, that it is at an angle to minimize reflection and that the seating is comfortable.

People often complain of eye problems when working with VDU's, but despite extensive research there is no evidence that they cause permanent damage or disease. However because the eyes are used so much in VDU work, people often become more aware of a problem that already existed. If they don't take frequent short breaks their eyes will become very tired and uncomfortable. Making sure that the VDU is positioned and adjusted to suit each individual, and that there is plenty of light, will improve their comfort. Contact lens wearers will often find that they need to use drops, or blink more because the heat generated by the equipment tends to dry the air. If this happens you will need to increase the humidity in the area. Bifocals are sometimes difficult to work with and it may be necessary to use different glasses for VDU work.

VDU's have been blamed for causing headaches, but again, there is no evidence that they, alone, are the cause. It is much more likely that glare from the screen or poor image quality is to blame; that the person needs different glasses; that they are worried about coping with the technology, the pace or the volume of work; that they are not comfortably seated or that they are not taking enough breaks. One, or several of these factors can cause headache.

There is no need to check radiation levels from a VDU or to wear any special protective equipment as VDU's do not give out harmful levels of radiation. They give out light and low-level electromagnetic radiation. Studies show that there is no link between the use of VDU's and miscarriage or birth defects so it is safe to continue working with them when you are pregnant.

Occasionally people suffer from rashes or irritation of the skin and this is probably caused by dry air and static electricity in a few people. More fresh air and higher humidity in the working environment can help.

In the main, VDU's do not affect people with epilepsy, although occasionally people with photosensitive epilepsy have been known to have problems. Usually people with epilepsy can work with VDU's.

6.1.2 How the Regulations affect you

The law says that employers have to minimize the risks in VDU work by making sure that workplaces and jobs are well designed; so what exactly does this mean?

The people who are affected by this are the ones who use a VDU as a major part of their everyday work. It applies to self-employed people too, because if you are using a client-employer's workstation, he or she has to assess and reduce risks, provide information and make sure it complies with the minimum requirements. If you work at home, this still applies.

Chapter 6

You may come across the terms 'user' and 'operator' A *user* is an employee and an *operator* is a self-employed person, both of whom habitually use display screen equipment as a significant part of their normal work. Both would be people to whom most or all of the following apply. A person:

- who depends on the display screen equipment to do their job;
- who has no discretion as to use or non-use;
- who needs particular training and/or skills in the use of display screen equipment to do their job;
- who uses display screen equipment for continuous spells of an hour or more at a time;
- who does so on a more or less daily basis;
- for whom fast transfer of data is important for the job;
- of whom a high level of attention and concentration is required, in particular to prevent critical errors.

If you are an employer, you need to (Fig. 6.1):

- analyze workstations and assess and reduce risks;
- make sure the workstations meet minimum requirements;

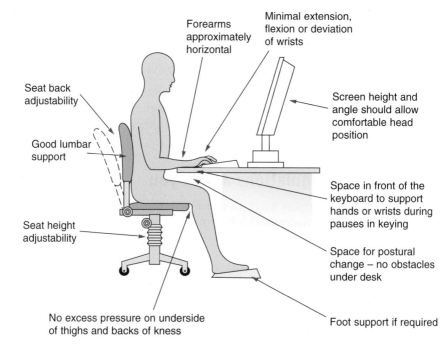

Figure 6.1 Workstation design.

◆ plan work to include changes of activity and breaks;
◆ arrange eye tests on request;
◆ provide special spectacles if needed; and
◆ provide training and information.

6.1.3 Analyzing the workstation

This involves looking at the equipment, furniture and work environment, the job itself and any special needs the staff may have.

Adjustable chairs and good lighting are typical **minimum requirements** and there is a schedule to the Regulations that you can look at to make sure that you are covering all these points.

The Regulations do not specify the frequency or length of breaks but **planning changes and breaks** will depend on the pressure and intensity of the work. Frequent short breaks are usually better. Ideally, staff should have some say over when they take a break.

If they are covered by the Regulations, employees can **request an eye test** at regular intervals, to be decided by the optometrist or doctor. The employer has to arrange and pay for this and also for any **special spectacles** that may be needed.

6.1.4 Training and information

Training and information about VDU health and safety must be provided by the employer. The Health and Safety Executive (HSE) publishes a straightforward and comprehensive guide called 'Working with VDU's' that will give valuable background information to employees. Employees also need to know how to use the equipment safely and how best to look after their health while using it.

6.2 Musculoskeletal disorders

6.2.1 Introduction

The term Musculoskeletal disorders refers to a wide range of conditions and injuries. These are broadly divided into two types:

1. Manual handling related disorders, such as backache, slipped discs, and sciatica; and
2. Work Related Upper Limb Disorders (WRULD) such as RSI, Tennis Elbow and Writers' Cramp.

Chapter 6

It has been estimated that some 600,000 people each year develop musculoskeletal disorders at a cost to employers of £1.25 billion. Apart from the pain and suffering to employees you must consider the losses which your business could suffer through lost time and possible litigation, as well as your legal duty to reduce the risks of such injury so far as is reasonably practicable.

Upper limbs means, the neck, shoulders, arms, wrists, hands and fingers. Upper limb disorders (sometimes called RSI) can happen in almost any workplace where people do repetitive, or forceful manual activities in awkward postures, for prolonged periods of time. These can cause muscular aches and pains which may be temporary at first, but if they are not properly controlled and the early symptoms are not recognized and treated these can become serious and disabling (Fig. 6.2).

Figure 6.2 Poor workstation layout may cause WRULD.

Manual handling including avoiding manual lifting injuries is covered in Section 4.6. This section deals mainly with WRULDS.

6.2.2 Decide if you have a problem

Employees with the following symptoms may have a problem:

- ◆ Aches and pains
- ◆ Swelling, numbness and tingling in hands, fingers, etc.
- ◆ Stiffness in joints
- ◆ Difficulty in movement.

Manual handling injuries mainly affect the back, arms, fingers and legs.

WRULDs mainly show themselves in the hands, wrists, arms and shoulders.

Check whether there is a previous history of upper limb disorder within your business.

6.2.3 Address the problem

- Avoid Manual handling operations, and risk assess those which cannot be avoided.
- Carry out a risk assessment of WRULD.

Risk factors to look for when assessing tasks in relation to upper limb disorders are:

- Activities which require a lot of force.
- The need for rapid, awkward or frequent movement.
- Awkward or static postures.
- Work for long periods without breaks or changes of activity.
- No special arrangements for new starters, or those returning to work after a long break.
- Poor environmental conditions, particularly vibration and low temperature.

There is some evidence that WRULDs increase where the speed, intensity or duration of the job increases. Employees may feel under stress when carrying out highly detailed work especially where there is lack of control over its speed and organisation, poor supervisory style or lack of support from management. In circumstances like these, people respond in two ways:

1. Physically (e.g. muscle tension); and
2. Behaviourally (e.g. adopting poor work methods or taking fewer rest breaks).

6.2.4 Implement your solutions

To reduce the risks to your employees of developing musculoskeletal disorders, you can take a number of approaches. Here are some examples,

Chapter 6

but you will need to look at your own specific work environment and tasks and decide what suits your business:

- ◆ Redesign the workplace
    - ○ to improve posture so that extreme joint postures are avoided
    - ○ to ensure a comfortable thermal environment
    - ○ to provide stable fixtures/jigs for assembly tasks
- ◆ Redesign the task
    - ○ to reduce the load or force needed
    - ○ to provide appropriate easy to handle tools
    - ○ to remove or reduce vibration
    - ○ to vary the tasks and the type of stresses they impose
- ◆ Work on product development
    - ○ reducing the number and complexity of fasteners and other components needed for assembly
    - ○ redesigning packaging for easier packing
    - ○ improving the hand holds and materials being used for example in lifting slings for care homes (Fig. 6.3)
    - ○ reducing the size and weight of the product and or components

 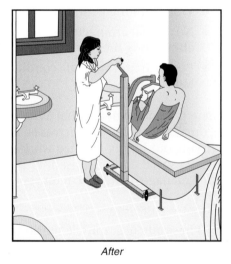

Before   *After*

Figure 6.3 Carehome.

- ◆ Look at how the work is organised
    - ○ avoid pacing and set work rates which are achievable across a shift
    - ○ consider effects of payment system on work stress or behaviour

- o provide breaks away from the workstation
- o provide training to improve task methods and workstation design
- o consider work rates for new employees and those returning – allow people to develop task fitness
- o Be careful when temporary manual methods are used following mechanical handling breakdowns for example lifting heavy paint tins onto pallets when the palletising machine breaks down, or reverting to the manual lifting of patients when the mechanical slings are being used for other patients.

6.2.5 Monitor the effects

You will have to check that the measures you are taking are effective, otherwise there is no point in investing in them.

- ◆ Have you achieved any targets you have set?
- ◆ Are sickness absence and employee complaints reducing?
- ◆ Is new equipment working properly and being maintained?
- ◆ Are employees using new tools and methods, not the old ones?
- ◆ Is training and refresher training up to date?

Managing musculoskeletal disorders is a process of continuous improvement. You will need to review and revise your assessments on a regular basis as your business changes and develops. Tackling these issues will improve the health of your employees and increase the efficiency and productivity of your business, so it will be well worth it in the long run.

Chapter 6

6.2.6 If you need further help?

- ◆ Consult your Trade Association
- ◆ Employ an ergonomist
- ◆ Consult an Occupational Health provider
- ◆ Approach your local enforcing authority
- ◆ Contact the Employment Medical Advisory Service (EMAS) through your local area HSE.

☞ *The following for further guidance Seat* IND(G)242(L)
☞ *Checkouts and Musculoskeletal Disorders* IND(G) 269.

6.3 Noise

6.3.1 Introduction

People's hearing can be irreversibly damaged by high levels of noise at work. These speed up the normal hearing loss that occurs as people grow older. Hearing loss caused by noise at work can be temporary or permanent. People often experience temporary deafness after leaving a noisy place and although hearing recovers within a few hours, this should not be ignored. It is a sign that if exposure to the noise is continued, your hearing could be permanently damaged. Sudden, extremely loud, explosive noises, for example, from guns or cartridge-operated machines can cause instantaneous permanent damage.

The length of time that a person is exposed to noise is also significant. Hearing loss is usually gradual because of prolonged exposure but it may only be when damage caused by noise over the years combines with hearing loss due to ageing that people realise how deaf they have become. Often this starts with the family complaining about the television being too loud, or maybe they find they cannot keep up with conversations in a group or they have trouble using the telephone. Eventually everything becomes muffled and people find it difficult to catch sounds like 't', 'd' and 's', so they confuse similar words.

As well as hearing loss, people may develop tinnitus (ringing, whistling, buzzing or humming in the ears), a distressing condition which can lead to disturbed sleep.

Remember: Young people can be damaged as easily as older people.

6.3.2 Regulations

The Control of Noise at Work Regulations 2005 requires employers to prevent or reduce risks to health and safety from exposure to noise at work. Employees also have duties under the Regulations.

The following list tells you what you as an employer must do:

- ◆ assess the risks to your employees from noise at work;
- ◆ take action to reduce the noise exposure to employees;
- ◆ provide your employees with hearing protection if you cannot reduce the noise exposure enough by using other methods;
- ◆ make sure the legal limits on noise exposure are not exceeded;

- provide your employees with information, instruction and training on the effects of exposure to noise and how to protect their hearing; and
- carry out health surveillance where there is a risk to health.

The Regulations do not apply to:

- people who are exposed to noise from their non-work activities, or to people who choose to go to noisy places; and
- low-level noise which is a nuisance but causes no risk of hearing damage.

Employers in the music and entertainment sectors have until 6 April 2008 to comply with the Control of Noise at work Regulations 2005. Meanwhile they must continue to comply with the Noise at Work Regulations 1989, which the 2005 Regulations replace for all other workplaces.

6.3.3 Is noise at work a problem?

Ask yourself two questions:

1. How loud is the noise?
2. How long are people exposed to it?

If any of the following apply, you will probably need to do something about it.

- Is the noise intrusive – like a crowded restaurant, a vacuum cleaner or a busy street – for most of the working day?
- Do your employees have to raise their voices to carry out a normal conversation when about 2 m apart for at least part of the day?
- Are noisy powered tools or machinery used by your employees for more than half an hour a day?
- Do you work in a noisy industry, for example construction, demolition or road repair; heavy vehicle operations; woodworking; plastics processing; engineering; textile manufacture; general fabrication; forging, pressing or stamping; paper or board making; canning or bottling; foundries?
- Are there noises due to impacts (such as hammering, drop forging, pneumatic impact tools, etc), explosive sources such as cartridge-operated tools or detonators, or guns?

Chapter 6

Consider too, that noise can be a safety hazard at work, because it can interfere with communication and make it difficult to hear machines or other danger approaching.

6.3.4 Measuring noise

Noise is measured in decibels (dB). An 'A-weighting' sometimes written as 'dB(A)', is used to measure average noise levels because it relates more to the way humans hear noise. A 'C-weighting' or 'dB(C)', is often used to measure peak, impact or explosive noises.

Because of the way our ears work, you might just notice a 3 dB change in noise level. However every 3 dB **doubles** the noise, so although the differences in the numbers might seem small, they can really be quite significant.

Some examples of typical noise levels are shown in Figure 6.4. This shows that a quiet office may range from 40–50 dB, while a road drill can produce 100–110 dB.

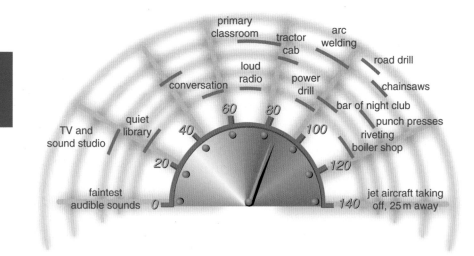

Figure 6.4 Examples of typical noise levels. *Source*: HSE from noise at work.

6.3.5 Risk assessment – Where do I start?

Section 6.3.3 asked 'Is noise a problem at work?' If you answered 'yes' to any of those questions you will need to decide whether any further action is needed, and plan how to do it by making a risk assessment.

The risk assessment will help you to decide what you need to do. It is more than just taking measurements of noise – in fact they may not even be necessary.

This is what you need to find out:

◆ Who is likely to be affected by noise in your business?

◆ Where is the risk of noise likely to be?

◆ You will need a reliable estimate of your employees' exposures to noise; how does this compare with the exposure action values (EAV) and limit values in Section 6.3.6?

◆ Find out what to do to comply with the law, for example are noise-control measures or hearing protection needed? Where are they needed? How much and what type do you need?

◆ Should any employees be provided with health surveillance? Are any at particular risk?

You must be able to show that your estimate of employees' exposure is representative of the work that they do. You need to look at:

◆ the tasks they are doing or are likely to do;

◆ how they carry out the tasks; and

◆ how the tasks may vary from day to day.

The information that you use to get your estimate must be reliable. This means using measurements in your own workplace, similar workplaces or information from suppliers of machinery.

The findings of your risk assessment need to be recorded. Make an action plan, and in it record anything you identify as being necessary to comply with the law. Set out what you have done and what you are going to do and make a timetable with action points assigned to named people.

If circumstances in your workplace change or you move elsewhere and noise exposures are affected, you will need to review your risk assessment. You should not leave it for more than 2 years before reviewing the risk assessment, even if it appears that nothing has changed.

6.3.6 What are 'action levels' and 'limit values'?

The Noise Regulations require you to take specific action at certain 'exposure action values' (EAV's). These relate to:

◆ the levels of exposure to noise of your employees averaged over a working day or week; and

◆ the maximum noise (peak sound pressure) to which employees are exposed in a working day.

The values are:

◆ lower EAVs:
  ○ daily or weekly exposure of 80 dB;
  ○ peak sound pressure of 135 dB;
◆ upper EAVs:
  ○ daily or weekly exposure of 85 dB;
  ○ peak sound pressure of 137 dB.

Figure 6.4 will help you decide what you need to do.

There are also noise exposure levels which must not be exceeded:

◆ exposure limit values (ELVs):
  ○ daily or weekly exposure of 87 dB;
  ○ peak sound pressure of 140 dB.

Any reduction in exposure provided by hearing protection which is worn is taken into account by these ELVs.

6.3.7 Noise – controlling the risks

Concentrate your efforts on controlling noise risks and noise exposure. This is the best way of ensuring that people do not lose their hearing.

The best route to controlling the risk from noise is to look for alternatives. If you can find equipment and/or working methods and processes

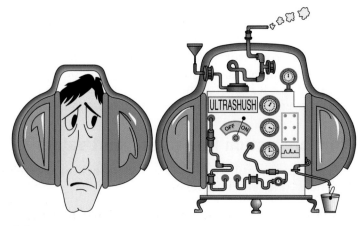

Figure 6.5 Control machine noise rather than wear hearing protection.

which would make the work quieter or mean people are exposed for shorter times, it is a better solution than providing hearing protection. You should also be keeping up with what is good practice or the standard for noise control within your industry (Fig. 6.5).

There are usually things you can do to reduce risks from noise, and if they are reasonably practicable, they should be carried out. But where noise exposures are below the lower EAVs you would only be expected to take actions which are not expensive and simple to carry out.

If your assessment shows that your employees are likely to be exposed at or above the upper EAVs, you must put in place a planned programme of noise control.

6.3.8 Risk assessment – using the information

Having produced a risk assessment, you are now equipped with the information you need on the risks and you will have an action plan for controlling noise. You can now:

- tackle the immediate risk, for example by reducing the time spent at a task or providing hearing protection;
- find out how you can control noise, how much it could sensibly be reduced;
- make priorities for action and put together a plan. Think, for example, what could be done immediately, and what requires longer term changes. Work out how these could be put in place. Calculate how many people would be exposed to the noise in each case?
- responsibilities need to be given to named people to carry out the various tasks both short term and long term including any noise control work;
- allocate sufficient resources to achieve the plan; and
- make sure that your action plan actually reduces the noise exposure.

6.3.9 Reducing noise

The best approach to reducing noise and noise exposure is often a combination of methods. Firstly, consider about how to remove the loud noise altogether, but if that cannot be done, put your efforts into controlling the noise at source. You may need to redesign the workplace and reorganise work patterns to reduce individual exposures. Take measures to protect individual workers if necessary.

Here are some ideas:

- ◆ Change the process or the equipment to
  - o do the work in a quieter way?
  - o replace whatever is causing the noise with something that is less noisy;
  - o introduce a low-noise purchasing policy for machinery and equipment;

- ◆ Limit the time spent in noisy areas – every time you halve the time spent in a noisy area it will reduce noise exposure by 3 dB;

- ◆ Introduce engineering controls:
  - o avoid metal-on-metal impacts, for example line chutes with abrasion-resistant rubber, and reduce drop heights;
  - o maintain your machinery properly and regularly as it deteriorates with age and will become noisier. Changes in noise levels can show that it is time to replace worn or faulty parts;
  - o fit noise insulating materials to partitions or enclosures to absorb the noise (thermal insulation is not suitable);
  - o anti-vibration mounts or flexible couplings can be used to isolate vibrating machinery or components from their surroundings;
  - o add material to reduce vibration (damping). Vibrating panels can be a source of noise;
  - o silencers to engines, air exhausts and blowing nozzles can be fitted.

- ◆ Change the paths by which the noise travels through the air to the people exposed, by, for example:
  - o erecting enclosures around machines to reduce the amount of noise emitted into the workplace or environment;
  - o using barriers and screens to block the direct path of sound;
  - o positioning noise sources further away from workers.

- ◆ Design and lay out the workplace for low noise emission, for example:
  - o reflected sound can be reduced within the building by using absorptive materials such as textiles, open cell foam or mineral wool;
  - o noisy machinery and processes can be kept away from quieter areas;
  - o identify areas where people spend most of their time and design the workflow to keep noisy machinery out of those areas;
  - o introduce quiet refuges for people to go to while minding equipment or doing paperwork.

6.3.10 Using hearing protection

Employees will need hearing protection:

◆ where extra protection is needed above what can been achieved using noise control;
◆ as a short-term measure while other methods of controlling noise are being developed.

Hearing protection should not be used as an alternative to controlling noise by technical and organisational means.

Use HSE's pocket card 'Protect your hearing or lose it!' Give this to your employees to remind them to wear their hearing protection.

6.3.11 The law and hearing protectors – What you need to do

◆ If noise exposure is between the lower and upper EAVs and your employees ask for it, you must provide them with hearing protectors.
◆ When noise exposure exceeds the upper EAVs, you must provide your employees with hearing protectors and make sure they use them properly.
◆ You need to identify hearing protection zones. These are areas where the use of hearing protection is compulsory. Mark them with signs if possible.
◆ Your employees must be provided with training and information on how to use and care for the hearing protectors;
◆ Make sure that the hearing protectors are properly used and maintained (Fig. 6.6).

6.3.12 Using hearing protection

Do's

◆ Aim at least to get below 85 dB at the ear. You need to make sure the protectors give enough protection.
◆ Target the use of protectors to the noisy tasks and jobs in a working day;

◆ Protectors should be chosen which are suitable for the working environment. How comfortable are they? Are they hygienic?

◆ If you are using other protective equipment such as hard hats, dust masks and eye protection, think about how hearing protectors will be worn (Fig. 6.6).

The problems of fitting earmuffs (e.g. with long hair, safety glasses or jewellery)

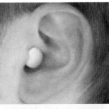

Correct

✓

The correct and incorrect fitting of earplugs

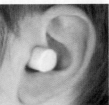

✗
Incorrect

Figure 6.6 Wearing hearing protection.

◆ Employees need to be able to choose items which suit them, so you will need to provide a range of protectors.

Don'ts

◆ Don't provide protectors which cut out too much noise – people can feel isolated, or not want to wear them.

◆ If the law doesn't require it, don't make the use of hearing protectors compulsory.

◆ Don't have a 'blanket' approach to hearing protection – it is much better to be more focused and only ask people to wear them when they need to.

6.3.13 Making checks

Here are some ideas to help you make sure that your employees use hearing protection when they are required to:

- ◆ Make sure the person who issues the protection is someone with respect and authority and replacements are readily available.
- ◆ Your safety policy should include the need to wear hearing protection.
- ◆ Make sure that hearing protection is being used properly and that the rules are being followed – this can be done by doing spot checks.
- ◆ you should follow your normal company disciplinary procedures where employees are not using protection properly.
- ◆ Limit the time that people spend in hearing protection zones. Make sure that only people who need to be there enter them, and that they do not stay longer than they need to.
- ◆ All managers and supervisors will need to be aware that they must set a good example by wearing hearing protection at all times when in hearing protection zones.

6.3.14 Maintenance of hearing protection

Hearing protection must work effectively. Check that:

- ◆ the equipment is kept clean and in good condition;
- ◆ there is no damage to earmuff seals;
- ◆ headband tension on earmuffs is still ok;
- ◆ no unofficial modifications have been made; and
- ◆ compressible earplugs are soft, pliable and clean.

6.3.15 Providing health surveillance

Health surveillance (hearing checks) must be provided for all your employees who are likely to be regularly exposed above the upper EAVs, or are at risk for any reason, for example they are particularly sensitive to hearing damage or have suffered some problems already.

The aim of health surveillance is to:

- ◆ alert you to employees who might be suffering from signs of early loss of hearing;

Chapter 6

♦ give you the chance to do something to prevent the damage getting worse;

♦ check that control measures you have put in are working.

Your trade union safety representative, or employee representative and the employees concerned should be consulted before you introduce health surveillance. Your employees need to understand that the aim of health surveillance is to protect their hearing. Their understanding and co-operation is essential if health surveillance is to be effective.

Health surveillance needs to be conducted by people who know what they are doing and have been properly trained. It is most likely that it will need to be supervised by a qualified health professional.

6.3.16 Employees what they need to know?

Employees must understand the risks they may be exposed to. You should at least tell them when they are exposed above the lower EAVs:

♦ the amount of noise they are likely to be exposed to and the risk to their hearing that this level of noise creates;

♦ where and how people can obtain hearing protection;

♦ what you are doing to control risks and exposures;

♦ how they can report defects in hearing protection and noise-control equipment;

♦ their duties are under the Control of Noise at Work Regulations 2005;

♦ how they can minimize noise exposure, by for example the proper positioning and use of noise-control equipment; how to look after and store hearing protection; and

♦ what health surveillance arrangements you have.

The information needs to be given in a way the employee can be expected to understand.

☞ *Noise at work a guide for employers* INDG362(rev1)

☞ *Protect your hearing or loose it!* INDG363(rev1)

6.4 Stress

Some people argue about whether stress is a real issue at work or whether it even exists, but research has shown that there is a link between poor working conditions and arrangements and ill health. The difference between 'stress' and 'pressure' needs to be understood in this context, because we experience pressure in the normal course of daily living; it's what motivates us and helps us to achieve; but 'stress' is too much pressure sustained over a long period so that the person doesn't have an opportunity to recover. The HSE defines stress as 'the adverse reaction people have to excessive pressure or other types of demand placed on them'. Stress at work should be controlled in the same way as other risks to health, by identifying the hazard, assessing who is at risk and how much they are at risk, deciding how to tackle the problem and putting plans into action.

In recent years there have been a number of high profile cases in court where individuals who have suffered severely from stress have gained large compensation awards. This is likely to increase with modern trends for long hours and high pressure working.

A problem often experienced by employers is that people believe that by admitting to experiencing stress at work they will be seen as weak or less than capable. It is up to managers and employers to create an atmosphere in which it is possible for staff to talk openly about any stress related problems. They also need to be aware of the symptoms and observant of the people working for them, so that they are able to pick up any signs of stress in the workplace and make necessary changes. The following list gives you the typical symptoms to look for:

◆ poor concentration
◆ performance dip
◆ uncharacteristic errors
◆ tantrums
◆ anti-social behaviour
◆ emotional outbursts
◆ alcohol or drug abuse
◆ nervous habits

People themselves may also report having sleep difficulties or loss of appetite.

There are six key areas of risk and you need to identify which ones are likely to cause problems in your business, so that you can work to reduce the difficulties that lead to stress. These areas are the basis of the HSE's new Management Standards:

◆ the demands of the job
◆ the management of change

- ◆ the role of the employee in the organisation
- ◆ the support received from managers and co-workers
- ◆ relationships at work
- ◆ the employee's control over their work

These standards provide a way of monitoring the performance of your workplace and will enable you to make improvements. It's worth considering that stress can have physical effects such as heart disease, back pain, gastro-intestinal disturbances and various other minor illnesses, and the psychological effects can lead, over time, to anxiety and depression. All this adds up to reduced performance and loss of working time through illness. There are added benefits to dealing with excessive stress and you can expect increased productivity, lower staff turnover and fewer absences from work in a stress free work environment.

☞   *Tackling Stress* INDG406, 2005
☞   *Working Together to reduce stress at work,* MISC686 single copies free from HSE Books

6.5 Vibration

6.5.1 The Regulations

The Control of Vibration at Work Regulations 2005 defines two aspects of vibration:

1. The first is known as hand-arm vibration (HAV) which is mechanical vibration transmitted into the hand and arms during work. Petrol driven hedge trimmers, chain saws and brush cutters and air powered road breakers typically cause problems.
2. The second is Whole Body Vibration (WBV) which is vibration transmitted into the body through the supporting surface when seated or standing, during a work activity.

The Regulations require you to:

- ◆ assess the vibration risk to your employees;
- ◆ decide if they are likely to be exposed above the daily EAV and if they are:
  - ○ introduce a programme of controls to eliminate risk, or reduce exposure to as low a level as is reasonably practicable;

Chapter 6

- ○ provide health surveillance (regular health checks) to those employees who continue to be regularly exposed above the action value or otherwise continue to be at risk;
- ◆ decide if they are likely to be exposed above the daily ELV and if they are: take immediate action to reduce it;
- ◆ exposure below the limit value;
- ◆ provide information and training to employees on health risks and the actions you are taking to control those risks;
- ◆ consult your trade union safety representative or employee representative on your proposals to control risk and to provide health surveillance;
- ◆ keep a record of your risk assessment and control actions;
- ◆ keep health records for employees under health surveillance; and
- ◆ review and update your risk assessment regularly.

6.5.2 Exposure Action Value (EAV) and Exposure Limit Value (ELV)

(a) Exposure Action Value (EAV)

The EAV is the amount of daily exposure to vibration above which you are required to take action to reduce risk.

Hand Arm Vibration (HAV)

For HAV it is set at $2.5\,m/s^2A(8)$ (metres/second^2 – A(8) denotes an average over an 8 hour day).

Whole Body Vibration (WBV)

For WBV it is set at a daily exposure of $0.5\,m/s^2A(8)$.

Whole-body vibration risks are low for exposures around the action value and only simple control measures are usually necessary in these circumstances.

(b) Exposure Limit Value (ELV)

The ELV is the maximum amount of vibration an employee may be exposed to on any single day.

Hand Arm Vibration (HAV)

For HAV it is set at $5\,m/s^2A(8)$.

The Regulations allow a transitional period for the limit value until July 2010 for equipment already in use before July 2007.

Whole Body Vibration (HBV)

For WBV it is set at a daily exposure of $1.15\,m/s^2A(8)$.

Chapter 6

The Regulations allow a transitional period for the limit value until July 2010 (or until 2014 for the agricultural and forestry sectors). This only applies to machines or vehicles first supplied to employees before July 2007. The ELV may be exceeded during the transitional periods as long as you have complied with all the other requirements of the Regulations and taken all reasonably practicable actions to reduce exposure as much as you can.

WHAT INJURIES CAN HAV CAUSE?

Regular exposure to HAV can cause a range of permanent injuries to hands and arms, collectively known as hand-arm vibration syndrome (HAVS). The injuries can include damage to the:

▼ Blood circulatory system (e.g. vibration white finger);

▼ Sensory nerves;

▼ Muscles;

▼ Bones;

▼ Joints.

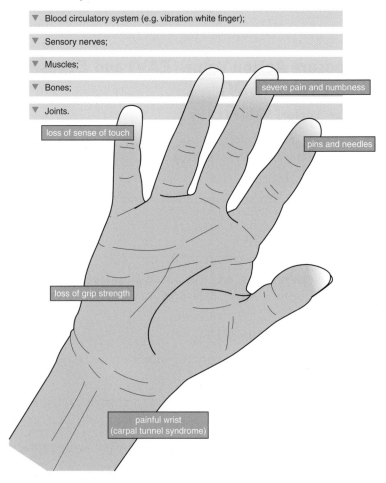

Figure 6.7 Injuries caused by HAV.

6.5.3 Effects on people

Excessive HAV exposure affects the nerves, blood vessels, muscles and joints of the hand, wrist and arm. It can become severely disabling if ignored and can cause 'vibration white finger' which can result in severe pain in the affected fingers and carpal tunnel syndrome which may involve pain, tingling, numbness and weakness in parts of the hand. WBV is likely to aggravate existing back pain or may in itself be the cause particularly when there is severe jolting or very high vibration (Fig. 6.7).

The effects on people include:

◆ pain, distress and sleep disturbance;
◆ inability to do fine work (e.g. assembling small components) or everyday tasks (e.g. fastening buttons);
◆ reduced ability to work in cold or damp conditions (i.e. most outdoor work) which would trigger painful finger blanching attacks; and
◆ reduced grip strength which might affect the ability to do work safely.

These effects can severely limit the jobs an affected person is able to do, as well as many family and social activities.

Most exposure to whole-body vibration at work is unlikely on its own to cause back pain. It may pose a risk when there is unusually high vibration or jolting or the vibration is uncomfortable for a long time on most working days. In such situations, the risk from vibration is related to the overall time the operator or driver is exposed to the vibration and the number of shocks and jolts they experience each day.

In some cases whole-body vibration can aggravate a back problem caused by another activity, for example a muscle strain caused by an accident when lifting a heavy object or during physical activity such as sport.

6.5.4 What jobs and equipment are likely to involve HAV and WBV vibration?

Jobs requiring regular and frequent use of vibrating tools and equipment and handling of vibrating materials are found in a wide range of industries, for example: building and maintenance of roads and railways; construction; estate management (e.g. maintenance of grounds, parks, water courses, road and railside verges); forestry; manufacturing concrete products; mines and quarries; and motor vehicle manufacture and repair.

There are hundreds of different types of hand-held power tools and equipment which can cause ill health from vibration. Some of the more common ones are: chainsaws; concrete breakers/road breakers; hammer drills; consolidation plates (whacker plates), hand-held grinders; impact wrenches; needle scalers; power hammers and chisels; powered sanders; strimmers/brush cutters (Fig. 6.8).

Figure 6.8 Road breaker may cause hand arm vibration problems.

Among those most likely to experience high WBV exposures are regular operators and drivers of off-road machinery such as: construction, mining and quarrying machines and vehicles, particularly earth-moving machines such as scrapers, vibrating rollers, bulldozers and building site dumpers; tractors and other agricultural and forestry machinery, particularly when used in transportation, tedding (turning hay), primary cultivation and mowing (Fig. 6.9).

The risk for road transport drivers from WBV exposure is likely to be low unless the vehicles do not have effective suspension (e.g. some types of smaller rigid-body lorries or flat-bed trucks) or are driven over poor surfaces or off-road. But there may be other causes of back pain for road transport drivers, which should probably be considered first, such as poor posture, long periods in the same position and repeated lifting and carrying.

6.5.5 Estimating exposure

It is a complex and expert job to measure HAV exposure but the HSE has suggested a simple points scheme in INDG175(rev2). This involves

Figure 6.9 Vibrating roller potential for whole body vibration.

taking the vibration information from the manufacturer or using Table 6.1 to estimate the figure.

You then use Table 6.2 to estimate the daily exposure.

For WBV you need to go to the manufacturer of the machine as many machine and vehicle activities in normal use will produce daily exposures below the ELV. But some off-road machinery operated for long periods in conditions that generate high levels of vibration or jolting may exceed the ELV. If you want to check you may be able to use the information in the vehicle manufacturer's handbook. You could also use data published by HSE and the exposure calculator on its website at www.hse.gov.uk/vibration for a range of machines and vehicles in different working conditions to make an estimate.

But it will be more effective for you concentrate your effort towards controlling the risks rather than trying to assess vibration exposures precisely.

6.5.6 Preventing vibration problems

To prevent HAV action should be taken to purchase machines with low vibration characteristics, maintain machinery properly, ensure cutting tools are kept sharp, ensure proper use of tools with correct grip and changes in the work pattern; store tools so that they do not have very cold handles, provide information and training for employees which includes keeping warm, exercising fingers and not smoking.

Table 6.1 Some typical vibration levels for common tools

Tool type	Lowest m/s²	Typical m/s²	Highest m/s²
Road breakers	5	12	20
Demolition hammers	8	15	25
Hammer drills/ combi hammers	6	9	25
Needle scalers	5	–	18
Scabblers (hammer type)	–	–	40
Angle grinders	4	–	8
Clay spades/jigger picks	–	16	–
Chipping hammers (metal)	–	18	–
Stone-working hammers	10	–	30
Chainsaws	–	6	–
Brush cutters	2	4	–
Sanders (random orbital)	–	7–10	–

Source: HSE

Table 6.2 Simple exposure points system

Tool vibration (m/s²)	3	4	5	6	7	10	12	15
Points per hour approx	20	30	50	70	100	200	300	450

Multiply the points assigned to the tool vibration by the number of hours of daily 'trigger time' for the tool and then compare the total with the EAV and ELV points.

100 points per day = Exposure action value (EAV)
400 points per day = Exposure limit value (ELV)

To reduce risk from WBV seats should be adjusted for the driver's weight, seat and control positions should be correctly adjusted, adjust vehicle speed to avoid excessive jolting, operate the machine smoothly, avoid travelling over rough terrain as far as possible, choose vehicles with the correct power and size for the work, keep roads and vehicles properly maintained.

☞ *Control the risks from hand-arm vibration* INDG175(rev 2) 2005

☞ *Control back pain risks from whole-body vibration* INDG242(rev10) 2005.

6.6 Violence and bullying

Violence at work causes pain, suffering, anxiety and stress, leading to financial costs due to absenteeism and higher insurance premiums to cover increased civil claims. It can be very costly to ignore the problem.

Violence at work is defined by the HSE as: 'any incident in which an employee is abused, threatened or assaulted in circumstances relating to their work'.

The HSE recommends the following four-point action plan:

6.6.1 Find out if there is a problem

You need to assess the risks and find out what the dangers are. You must ask people in the workplace and sometimes a short questionnaire will help you to get at the facts. Record all incidents to get a picture of what is happening over time, and make sure that every relevant detail is recorded.

6.6.2 Decide on what action to take

It is important to weigh up the risks and decide who is likely to suffer and in what way. The threats may be from the public or co-workers at the workplace or it may be as a result of visiting the homes of clients. Consult with your employees or with other people at risk. This way you will make sure that they are committed to any plans that you put together to prevent violence and bullying.

6.6.3 Take the appropriate action

The arrangements for dealing with violence should be included in the safety policy and managed like any other aspect of the health and safety procedures. Action plans should be drawn up and followed through using the consultation arrangements as appropriate. The police should also be consulted to ensure that they are happy with the plan and are prepared to play their part in providing back up, if it becomes necessary.

6.6.4 Check that the action is effective

Ensure that the records are being maintained and any reported incidents are investigated and suitable action taken. The procedures should be regularly audited and changes made if they are not working properly.

People who have suffered violence or bullying need help and assistance to overcome their distress Debriefing, counselling, time off to recover, legal advice and support from colleagues should be made available to them.

☞   *Violence at Work* INDG69(rev)

Appendix 6.1
Workstation self assessment checklist

Name: Department: Date:

The completion of this checklist will enable you to carry out a self-assessment of your own workstation. Your views are essential in order to enable us to achieve our objective of ensuring your comfort and safety at work. Please circle the answer that best describes your opinion, for each of the questions listed. The form should be returned to.................................. as soon as it has been completed.

Environment

1. *Lighting*			
Describe the lighting at your usual workstation.			
	about right	too bright	too dark
Do you get distracting reflections on your screen?			
	never	sometimes	constantly
What control do you have over local lighting?			
	full control	some control	no control
2. *Temperature and humidity*			
At your workstation, is it usually:			
	comfortable	too warm	too cold?
Is the air around your workstation:			
	comfortable		too dry?
3. *Noise*			
Are you distracted by noise from work equipment?			
	never	occasionally	constantly
4. *Space*			
Describe the amount of space around your workstation.			
	Adequate		Inadequate

Chapter 6

Furniture

5. Chair

Can you adjust the height of the seat?

        yes   no

Can you adjust the height and angle of the backrest?

        yes   no

Is the chair stable?

        yes   no

Does it allow movement?

        yes   no

Is the chair in a good state of repair?

        yes   no

If your chair has arms, do they get in the way?

        yes   no

6. Desk

Is the desk surface large enough to allow you to place all your equipment where you want it?

        yes   no

Is the height of the desk suitable?

        yes   too high   too low

Does the desk have a matt surface (non-reflectant)?

        yes   no

7. Footrest

If you cannot place your feet flat on the floor whilst keying, has a footrest been supplied?

        yes   no

8. Document holder

If it would be of benefit to use a document holder, has one been supplied?

        yes   no

If you have a document holder, is it adjustable to suit your needs?

        yes   no

Display screen equipment

9. Display screen

Can you easily adjust the brightness and the contrast between the characters on screen and the background?

        yes   no

Does the screen tilt and swivel freely?

        yes   no

Is the screen image stable and free from flicker?

              yes   no

Is the screen at a height, which is comfortable for you?

              yes   no

10. Keyboard

Is the keyboard separate from the screen?

              yes   no

Can you raise and lower the keyboard height?

              yes   no

Can you easily see the symbols on the keys?

              yes   no

Is there enough space to rest your hands in front of the keyboard?

              yes   no

11. Software

Do you understand how to use the software?

              yes   no

12. Training

Have you been trained in the use of your workstation?

              yes   no

Have you been trained in the use of software?

              yes   no

If you were to have a problem relating to display screen work, would you know the correct procedures to follow?

              yes   no

Do you understand the arrangements for eye and eyesight tests?

              yes   no

Any other comments?

Chapter 6

Construction and contractors

Summary

- Basics of contractor management
- Some of the hazards most commonly found on construction sites

7.1 Introduction

This chapter is for small contractors, sub-contractors or suppliers.

A **contractor** is anyone who is called in to work for a company but who is not an employee of that company. A **supplier** is someone who supplies goods or services.

Working safely with large firms can help your business – not working safely can damage your business – and theirs. What is more, it could get you, and them, a bad name!

It's your job to manage health and safety, and this chapter gives you basic information about your responsibilities and it also tells you about the most common requirements that clients are likely to ask you about when assessing your competence in managing health and safety.

Many organisations depend on a number of others (often firms smaller than themselves) in order to produce their goods on time and to a reliable standard. Good health and safety standards can make these relationships better for everyone.

7.2 Contractors

Contracting out some types of tasks, for example installation and maintenance of equipment, disposing of waste, cleaning premises, and painting and decorating is common practice in large firms (Fig. 7.1).

These tasks can be especially dangerous because they have to be done on clients' sites, and in situations which are unfamiliar to you, the contractor. Accidents can happen because you are unfamiliar with the dangers on site and often the people on site do not realise you are there, working close by. If properly assessed and managed such accidents can be avoided.

Engaging safe contractors is important for many companies. This starts with a thorough check of the contractor's competence before the contract is agreed. It should continue throughout the work via close co-operation and communication with everyone involved. An appropriate level of supervision is also needed. When the contract has been completed companies often review the health and safety performance of contractors and any sub-contractors and keep records for future reference. This is often in the form of an approved list of contractors. Getting on to an 'approved' list will depend on a number of factors such as price, competence, quality control and financial stability as well as your health and safety performance.

Figure 7.1 Painting contractor at work.

Of course, it is always in your interests to work safely, but if you want to work with other companies you will need to be able to explain how you manage health and safety.

7.3 Suppliers

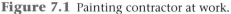

If you are a supplier then much of the information in this chapter will apply to you. Just as important to firms are the suppliers of components, particularly if you supply critical components which go directly on to an assembly line (Fig. 7.2).

Not being able to meet your client's needs for goods and services, because there has been an accident, or you have had a visit from a health and safety inspector who has taken enforcement action against you, may mean you lose the business. To avoid any such difficulties, clients may want to audit your business which can be a stressful experience.

Chapter 7

Figure 7.2 Plant company supplying lights for a construction site.

All parties to a contract have specific responsibilities under health and safety law, which cannot be passed on to other people. This is quite different from contract law where often the responsibility will depend on the wording of the contract. Here are the health and safety responsibilities:

♦ Employers must protect people from harm caused by their work activities. This includes the responsibility not to harm contractors and sub-contractors on their site.

♦ Employees must co-operate with their employer on health and safety matters, and not do anything that puts them or others at risk.

♦ Employers must train their employees and give them clear instructions about their duties. In some cases there are special legal requirements for competence – for example, a gas installer must be CORGI registered and someone stripping old asbestos lagging material would need to be licenced.

♦ If you are self-employed you must not put yourself in danger, or put others in danger who may be affected by what you do.

♦ Suppliers of chemicals, machinery and equipment have to make sure their products or imports are safe, and without risks to health. They must also provide information about the health and safety of their products.

7.4 What people need to know

Relationships can get quite complicated and it is therefore important to share information and agree what has to be done, how it is going to be done, who is in charge, and who is responsible. Here is a list of issues:

1. *Communicate with clients, other contractors and employees*
   - Before you start the work.
   - As you progress, and particularly if you find that you need to make significant changes in order to get the job done.

2. *Control and co-ordinate the work*
   - Make sure everyone knows who is in control and co-ordinating the work and the risks that might arise.
   - Ensure department heads, managers and supervisors understand their responsibilities;
   - Make certain that employees know what they must do, how they will be supervised and how they will be held accountable.

3. *Co-operate with others*
   - Co-operate with others to ensure that one contractor's work does not adversely affect the work of others.
   - Agree responsibilities in advance of the work.
   - consult your staff and their representatives and involve them in planning the work.

4. *Check out competence*
   - Make sure employees have the necessary skills, training and experience to carry out all tasks safely – double check this with people on especially dangerous work.
   - Provide employees with the relevant health and safety information, instruction and training for the job to be done.
   - Ask for advice and information where you need it.

7.4.1 Essential health and safety laws

The two most important health and safety laws for employers and contractors are:

- The Health and Safety at Work, etc. Act 1974.
- The Management of Health and Safety at Work Regulations 1999.

Chapter 7

A number of other regulations applicable to contractual situations which require risk assessments to be made are:

◆ The Construction (Design and Management) Regulations 2007 (CDM2007).
◆ The Construction (Head Protection) Regulations 1989.
◆ The Control of Asbestos Regulations 2006.
◆ The Control of Substances Hazardous to Health (COSHH) Regulations 2002.
◆ The Manual Handling Operations Regulations 1992.
◆ The Control of Noise at Work Regulations 2005.
◆ Personal Protective Equipment at Work Regulations 1992 (PPE).
◆ The Work at Height Regulations 2005.
◆ The Control of Vibration at Work Regulations 2005.

7.5 Construction and maintenance jobs

For many construction, refurbishment, maintenance and repair (including redecoration) works the CDM Regulations 2007 will apply. The CDM 2007 Regulations apply to all construction projects. Larger projects which are notifiable (i.e. last for more than 30 days or will involve more than 500 person days of work), have more extensive requirements. You should consult the 'Managing health and safety in construction, Approved Code of practice. L144 HSC 2007 HSE Books (Figs. 7.3 and 7.4).'

All projects require:

◆ Non-domestic clients to check the competence of all their appointees; ensure there are suitable management arrangements for the project; allow sufficient time and resources for all stages; provide pre-construction information to designers and contractors.
◆ Designers to eliminate hazards and reduce risks during design; and provide information about remaining risks.
◆ Contractors to plan, manage and monitor their own work and that of workers; check the competence of all their appointees and workers; train their own employees; provide information to their workers; comply with the requirements for health and safety on site detailed in Part 4 of the Regulations and other regulations such as the Work at Height Regulations; and ensure there are adequate welfare facilities for their workers.

Figure 7.3 Building site CDM applies to all parties.

Figure 7.4 Domestic clients do not have responsibilities under CDM2007 but the Regulations apply to the contractor.

◆ Everyone to assure their own competence; co-operate with others and coordinate work so as to ensure the health and safety of construction workers and others who may be affected by the work; report obvious risks; take account of the general principles of prevention in planning or carrying out construction work; and comply with the requirements in Schedule 3, Part 4 of CDM2007 and other regulations for any work under their control.

7.6 Providing a health and safety method statement

You may be asked by a client to prepare a detailed method statement describing how you intend to carry out the job including all the control measures which will be applied.

A method statement is a formal written document detailing how an activity or task is to be undertaken. It pays particular attention to the health, safety and welfare implications in carrying out such an activity or task.

As a general rule, if potentially hazardous activities are to be undertaken then method statements should be prepared. Health and safety method statements are not required by law, but they have proved to be an effective and practical management tool, especially for higher risk work. They are commonly required for construction and demolition work.

If the work is to be carried out by subcontractors then they should prepare and issue the method statement.

Typical work which will require method statements include:

(a) Erection and dismantling of design scaffolding, temporary support systems, form work, false work, etc.

(b) Demolition work.

(c) Excavation work below 1.2 metres (Fig. 7.5b).

(d) Refurbishment work which may affect the structural stability of such a structure.

(e) Roof work.

(f) Erection of structures.

(g) Work on high voltage electrical equipment.

(h) Entry into confined spaces (Fig. 7.5a).

(i) Hot work.

(j) Work involving highly flammable liquids such as petrol.

The method statement should be based on your assessment of the risks to the health and safety of employees and others who could be affected by the work. The findings of risk assessments should be incorporated into the method statement.

Chapter 7

Figure 7.5a Confined space – method statement needed.

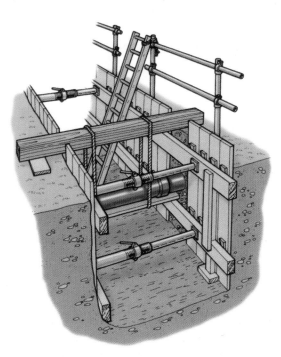

Figure 7.5b Excavation – method statement needed.

Chapter 7

The extent and detail of a method statement will depend upon the size and/or complexity of the work, activity or task to be undertaken. A method statement should contain the following:

1. Management arrangements, including identified persons with authority.
2. Detailed sequence of work operations in chronological order.
3. Drawings and/or technical information.
4. Detailed information on plant, equipment, substances, etc.
5. Inspection and monitoring controls.
6. Risk assessments.
7. Emergency procedures and systems
8. Arrangements for delivery, stacking, storing and movement of logistics on site.
9. Details of site features, layout and access, which may affect method of working.
10. Procedures for changing or departing from method statement.

It must be remembered that the method statement is a dynamic document and must be adhered to and kept up to date.

7.7 Subcontracting work

Client companies may have to satisfy themselves that you, as contractor, have an effective procedure for appraising the performance of a subcontractor. You can use some or all of the following criteria to do this appraisal. It is likely that clients will ask you the same kinds of questions about your own business. Questions to ask are:

- Do they have an adequate health and safety policy?
- Can they demonstrate that the person responsible for the work is competent?
- Can they demonstrate that competent safety advice will be available?
- Do they have any independent assessments of their competence?
- Do they monitor the level of accidents at their work site?
- What is their recent health and safety performance like? (How many accidents have they had and so on.)
- Do they have a system to assess the hazards of a job and implement appropriate control measures?

- Will they produce a method statement, which sets out how they will deal with all significant risks?
- Do they have guidance on health and safety arrangements and procedures to be followed?
- Do they have effective monitoring arrangements?
- Do they use trained and skilled staff who are qualified where appropriate? (Judgement will be required, as most construction workers have had no training except training on the job.) Can the company demonstrate that their employees or other workers used for the job have had the appropriate training and are properly qualified?
- Can they produce good references indicating satisfactory performance?
- How much experience do they have in the type of work you are asking them to do?
- What kind of arrangements do they have for consulting their workforce?
- If applicable, do they or their workers hold a 'passport' in health and safety training?
- How do they provide health and safety training and supervision for their employees?
- Are they members of a trade or professional association?

You will need to discuss proposed working methods with sub-contractors before giving them a contract. Find out how they are going to work, what equipment and facilities they are expecting to be provided and what equipment they will bring to the site.

Identify any health or safety risks which their operations may create for others working at the site. Agree control measures.

The information in this chapter should help you to assess the health and safety competence of sub-contractors.

7.8 Safety rules for contractors

In the conditions of contract there is often a stipulation that the contractor and all of their employees adhere to the contractors safety rules. Contractors safety rules should contain as a minimum the following points:

(a) *Health and safety*: That the contractor operates to at least the minimum legal standard and conforms to accepted industry good practice.

(b) *Supervision*: That the contractor provides a good standard of supervision of their own employees.

(c) *Sub-contractors*: That they may not use subcontractors without prior written agreement from the company.

(d) *Authorisation*: That each employee must carry an authorisation card issued by the company at all times whilst on site.

See Appendix 7.1 for details of safety rules.

7.9 Working with a permit-to-work system

You may be expected to be familiar with a 'permit-to-work system'. This is a formal written system used to control certain types of work that are potentially hazardous.

Examples include entry into vessels, hot work, pipeline breaking, and isolation. Permits-to-work form an essential part of safe systems of work for many maintenance activities, particularly in the chemical industry.

7.10 Construction hazards

Construction is a high-risk industry. The detailed health and safety requirements for construction sites are contained in CDM2007 Part 4 and other regulations, for example Work at Height Regulations. Here's the 'High 5' that will help you keep safe and healthy as suggested by the HSE.

7.10.1 High 1: the basics

Tidy sites and decent welfare

Tidy sites and decent welfare are the basics of a good site. Slips and trips are the most common cause of injuries at work.

An untidy site is a poorly managed site.

All sites need decent welfare facilities. The minimum welfare requirements are:

♦ Clean toilets.
♦ Running hot and cold water with soap and towels.

Figure 7.6 Typical site entrance notices.

- Basins large enough to immerse your arms up to the elbows.
- Drinking water.
- Somewhere warm, dry and clean to sit and eat.

Poor welfare facilities can lead to ill health (Fig. 7.7).

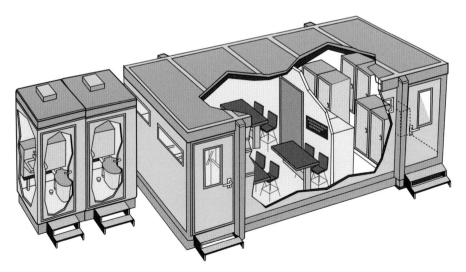

Chapter 7

Figure 7.7 Mobile site toilets and welfare facilities.

7.10.2 High 2: falls from height

Falls from height are the biggest cause of fatal and serious injuries during construction work. They account for 50% of all deaths. Many accidents involve falls from roofs, through fragile materials, from ladders and from leading edges (Fig. 7.8).

Figure 7.8 Scaffolding.

(a) *Generally make sure you*

- ◆ Work from a safe and secure place or platform with proper edge protection.
- ◆ Use scaffolds and scaffold towers that are competently erected.
- ◆ Use powered access equipment safely.
- ◆ Protect holes and leading edges, for example, with guardrails and toe boards.

(b) *When working on roofs never*

- ◆ Work in poor weather.
- ◆ Work on sloping roofs without edge protection.
- ◆ Throw down waste or equipment.

Chapter 7

Take care when working on or near fragile material – you can fall through as well as off it.

(c) *Ladders* (see Chapter 4 – work at night)
- ◆ Only use ladders for light work of short duration if there's no alternative.
- ◆ Angle and secure them to prevent slipping (1 out for 4 up) (Fig. 4.26).
- ◆ Always make sure ladders are properly maintained.
- ◆ Never overreach.

7.10.3 High 3: manual handling

Manual handling injuries from working with heavy, awkward materials, often in cold and wet conditions, are one of the most common reasons why workers leave construction. Injuries are made worse by repetitive jobs, such as laying heavy blocks.

Figure 7.9 Brick elevator to get materials safely to the top of a scaffold.

- ◆ Use mechanical means, for example, hoists, teleporters and chutes rather than hods (Fig. 7.9).
- ◆ Choose equipment suitable for the job and keep it maintained.
- ◆ Change to lighter materials, bags, etc.
- ◆ Avoid repetitive handling.
- ◆ Avoid awkward movements.
- ◆ Protect yourself and reduce the strain.

7.10.4 High 4: transport

Workplace transport incidents are the second most common cause of fatalities after falls from height.

- ◆ Use barriers and warning signs to separate vehicles and people (Fig. 7.10).
- ◆ Create clearance around slewing vehicles.

Figure 7.10 People and vehicles separated.

- Avoid reversing – where you can't, use trained banksmen.
- Make sure loads are secure.
- Don't use plant and vehicles on dangerous slopes.
- Only take passengers on vehicles designed to take them.
- Make sure vehicles are maintained and operators are trained.

When people and vehicles collide, people come off worse – so keep them apart!

7.10.5 High 5: asbestos

Many buildings in the UK contain asbestos. If you're thinking of working in a building that was built or renovated up to the 1980s you should assume that it contains asbestos until proved otherwise.

The main asbestos-containing materials (ACMs) are lagging, asbestos insulating board, decorative coatings and asbestos and asbestos cement.

- Check if there is any ACM.
- Find out what you need to do to work safely.
- If in doubt leave it to the experts.

See Chapter 5.2 for detailed information on asbestos.

Chapter 7

Appendix 7.1

Sample safety rules for contractors

Rules for contractors

(*Note*: This is written in a more formal style so that you could use it to write a formal agreement with a contractor.)

Contractors engaged by the Company to carry out work on its premises will:

- familiarise themselves with so much of the Company's Safety Policy as affects them and will ensure that appropriate parts of the Policy are communicated to their employees, and any sub-contractors and employees of sub-contractors who will do work on the premises;
- co-operate with the Company in its fulfilment of its health and safety duties to contractors and take the necessary steps to ensure the like co-operation of their employees;
- comply with their legal and moral health, safety and food hygiene duties;
- ensure the carrying out of their work on the Company's premises in such a manner as not to put either themselves or any other persons on or about the premises at risk;
- where they wish to avail themselves of the Company's first aid arrangements/ facilities whilst on the premises, ensure that written agreement to this effect is obtained prior to first commencement of work on the premises;
- where applicable and requested by the Company, supply a copy of its Statement of Policy, Organisation and Arrangements for health and safety written for the purposes of compliance with The Management of Health and Safety at Work Regulations 1999 and Section 2(3) of the Health and Safety at Work, etc. Act 1974;
- comply with all applicable requirements under the Construct (Design and Management) Regulations 2007;
- abide by all relevant provisions of the Company's Safety Policy including compliance with health and safety rules; and
- ensure that on arrival at the premises, they and any other persons who are to do work under the contract report to Reception or their designated Company contact.

Without prejudice to the requirements stated above, contractors, sub-contractors and employees of contractors and sub-contractors will, to the extent that such matters are within their control, ensure:

- the safe handling, storage and disposal of materials brought onto the premises;

Chapter 7

- that the Company is informed of any hazardous substances brought onto the premises and that the relevant parts of the Control of Substances Hazardous to Health Regulations in relation thereto are complied with;
- fire prevention and fire precaution measures are taken in the use of equipment which could cause fires;
- steps are taken to minimise noise and vibration produced by their equipment and activities;
- scaffolds, ladders and other such means of access where required are erected and used in accordance with Work at Height Regulations and good working practice;
- any welding or burning equipment brought onto the premises is in safe operating condition and used in accordance with all safety requirements;
- any lifting equipment brought onto the premises is adequate for the task and has been properly tested/certified;
- any plant and equipment brought onto the premises is in safe condition and used/operated by competent persons;
- for vehicles brought onto the premises, any speed, condition or parking restrictions are observed;
- compliance with the relevant requirements of the Electricity at Work Regulations 1989;
- connection(s) to the Company's electricity supply is from a point specified by its management and is by proper connectors and cables;
- they are familiar with emergency procedures existing on the premises;
- welfare facilities provided by the Company are treated with care and respect;
- access to restricted parts of the premises is observed and the requirements of Food Safety Legislation is complied with;
- any major or lost-time accident or dangerous occurrence on the Company's premises is reported as soon as possible to their site contact; and
- where any doubt exists regarding health and safety requirements, advice is sought from the site contact;

The foregoing requirements *do not* exempt contractors from their statutory duties in relation to health and safety, but are intended to assist them in attaining a high standard of compliance with those duties.

Chapter 7

Appendix 7.2

Health and Safety Checklist for contractors

1. Checklist

This checklist should remind you about the topics you might need to discuss with the people you will be working with.

It is not intended to be exhaustive and not all questions will apply at any one time, but it should help you to get started.

a. *Responsibilities*
- What are the hazards of the job?
- Who is to assess particular risks?
- Who will co-ordinate action?
- Who will monitor progress?

b. *The job*
- Where is it to be done?
- Who with?
- Who is in charge?
- How is the job to be done?
- What other work will be going on at the same time?
- How long will it take?
- What time of day or night?
- Do you need any permit to do the work?

c. *The hazards and risk assessments*
  i. Site and location

Consider the means of getting into and out of the site and the particular place of work – are they safe; and
- Will any risks arise from environmental conditions?
- Will you be remote from facilities and assistance?
- What about physical/structural conditions?
- What arrangements are there for security?

  ii. Substances
- What supplier's information is available?
- Is there likely to be any microbiological risk?

- What are the storage arrangements?
- What are the physical conditions at the point of use? Check ventilation, temperature, electrical installations, etc.;
- Will you encounter substances that are not supplied, but produced in the work, for example, fumes from hot work during dismantling plant? Check how much, how often, for how long, method of work, etc.
- What are the control measures? For example, consider preventing exposure, providing engineering controls, using personal protection (in that order of choice).
- Is any monitoring required?
- Is health surveillance necessary, for example, for work with sensitisers? (refer to health and safety data sheet)

iii. **Plant and equipment**

- What are the supplier/hirer/manufacturer's instructions?
- Are any certificates of examination and test needed?
- What arrangements have been made for inspection and maintenance?
- What arrangements are there for shared use?
- Are the electrics safe to use? Check the condition of power sockets, plugs, leads and equipment. (Don't use damaged items until they have been repaired.)
- What assessments have been made of noise levels?

d. People

- Is information, instruction and training given, as appropriate?
- What are the supervision arrangements?
- Are members of the public/inexperienced people involved?
- Have any disabilities/medical conditions been considered?

e. Emergencies

- What arrangements are there for warning systems in case of fire and other emergencies?
- What arrangements have been made for fire/emergency drills?
- What provision has been made for first aid and fire-fighting equipment?
- Do you know where your nearest fire exits are?
- What are the accident reporting arrangements?

Chapter 7

- Are the necessary arrangements made for availability of rescue equipment and rescuers?

f. Welfare

Who will provide:

- shelter;
- food and drinks;
- washing facilities;
- toilets (male and female); and
- clothes changing/drying facilities?

You may be faced with other pressing requirements – but re-think health and safety as the work progresses.

Accidents and emergencies

8

Summary

- Accidents including the causes, investigating and reporting requirements
- Planning for emergencies
- First aid and emergency procedures
- Form for accident investigations
- Roles and powers of enforcing officers
- Injury claims for compensation

8.1 Introduction

The chapter addresses you as an employer but it also applies to self-employed people, or people who are in control of premises.

The aim of health and safety management is to prevent accidents from happening by ensuring that all safety precautions, controls and procedures are in operation and working. However, in any organisation, things sometimes go wrong and accidents can still happen. Procedures need to be set up to deal with these unplanned events. Managers and employees need to think about accident and emergency situations before they happen and plan how to deal with them.

What is an accident? An accident is an unplanned, unpremeditated event that results in personal injury and/or property damage or could have resulted in personal injury and/or damage to equipment and/or other losses to an organisation.

Accidents can cause suffering, pain, distress, serious injury and disability. Accidents can also result in lost production time, medical costs, compensation proceedings, damage to property and equipment and paperwork.

8.2 Accidents can cost a great deal

Minor injuries cost more than the price of a plaster. This is because of other costs such as:

- ◆ the first-aider's time;
- ◆ the injured person's absence from work, particularly if a trip to the hospital is needed;
- ◆ finding another worker to replace the injured person;
- ◆ people stopping work to help or being absent from work with stress related illnesses following the accident; and
- ◆ loss of customer confidence if it becomes known that a worker has been injured (Fig. 8.1).

Consider how the costs of these apparently minor accidents can add up over a year, especially if people have to take time off work.

Accidents involving plant, equipment and production can cost even more.

The total accident bill will depend on the number of employees involved, the type of work and the value of raw materials, products or services.

Figure 8.1 Ladder accident.

Over a year the total cost can be significant. Sadly, in some cases a single accident can mean the end of a business.

Accidents can also result in increased insurance premiums. There are consequences of accidents that cannot easily be calculated – damage to your company's image, loss of business, effects on employee morale, reduced efficiency, fines and even, in some cases, imprisonment.

8.3 Causes of accidents

There are many causes of accidents at work both human and organisational. Some are immediate causes while others are root or underlying causes. Here are the most common to think about:

Immediate causes – personal factors

- ◆ Behaviour of the people involved.
- ◆ Ignorance of risks.
- ◆ Suitability of the people doing the work.
- ◆ Training and competence.
- ◆ Lack of safety equipment/clothing.
- ◆ Poor safety awareness.

Chapter 8

◆ Tiredness.
◆ Stupidity/fooling around.

Immediate causes – task factors

◆ Poor housekeeping/untidiness.
◆ Work place precautions and controls.
◆ Faulty/unprotected equipment.
◆ Badly designed equipment.
◆ Uncomfortable working environment, for example, too noisy, poor lighting, too cold, etc.
◆ Failure to use appropriate safety signs or warnings.
◆ Lack of a safe system of work (Figs. 8.2 and 8.3).

Figure 8.2 Poor working environment.

Root or underlying causes

◆ Low standards of information, communications, instruction, training and supervision.
◆ Previous similar incidents.
◆ Lack of control and co-ordination of the work.
◆ Quality of co-operation and consultation with employees.
◆ Quality of health and safety policy and procedures.

Figure 8.3 Poor scaffold.

- Deficiencies in risk assessments, plans and control systems.
- Deficiency in monitoring and measurement of work activities.
- Quality and frequency of reviews and audits.

8.4 Emergency procedures

When things go wrong people are often exposed to danger. There should be an emergency plan to deal with major incidents or serious injuries.

As an employer you need to think about the following:

- What is the worst that could happen?
- How would you deal with the situation? Who would be in charge? Do they need training? Competent persons should be nominated to take control. It is important to discuss this and liaise with the emergency services.
- Could emergency services gain easy access to the premises?

Chapter 8

◆ How should the alarm be raised, both when the premises are open and closed?

◆ Where would people go to reach a place of safety, to get help or safety equipment?

◆ Are there enough emergency exits for a quick escape?

◆ Are emergency doors checked regularly?

◆ Are escape routes unobstructed and clearly marked with running persons symbols and arrows indicating the direction of escape?

◆ Who are the other key people such as first-aiders?

◆ How to plan essential actions such as emergency plant shut-down or making processes safe.

◆ Training everyone in emergency procedures (Fig. 8.4).

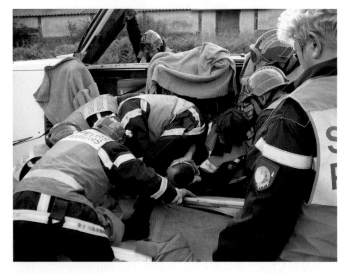

Figure 8.4 Emergency training exercise.

In the event of a fire

Here is a model evacuation procedure. This is intended as a guide only and will need to be adapted according to the needs of your business.

1. Raise or sound the alarm.

2. Call for assistance and attack the fire with the correct fire extinguishing equipment provided, if it is safe to do so.

3. Call the fire brigade by dialling 999.

4. Leave the premises and report immediately to the assembly area. Do not stop to collect personal property. Do not use a lift.

5. Stay at the assembly point until otherwise instructed. It will be necessary to take a roll call to account for everyone present.

6. Warn others in the building or nearby that fire has broken out.

7. NEVER underestimate a fire.

An example of an emergency procedure for a shop premises is shown in Appendix 8.2 and a fire notice at Appendix 8.3.

8.5 Investigating accidents and incidents

ALL accidents should be investigated – not only accidents that cause injury, but also incidents in which no injury was sustained, but where there was damage to plant or equipment.

It is important to investigate accidents:

- ◆ to prevent a recurrence;
- ◆ in order to report them to the authorities as required; and
- ◆ to record the facts for future reference and analysis.

When an accident happens you should:

- ◆ take any action required to deal with the immediate risks;
- ◆ think about what kind of investigation is needed;
- ◆ make sure that the accident is properly recorded in the accident book;
- ◆ report the accident as required to the enforcing authorities (see Section 8.7);
- ◆ investigate – find out what happened and why; and
- ◆ take action to stop something similar happening again.

Also look at near misses and property damage. Often it is only by chance that someone wasn't injured. Any investigation should discover:

- ◆ Details of anyone who was injured, details of the injury, damage or loss to property or equipment.
- ◆ What happened, where, when, what was the cause, were there any underlying causes? (Fig. 8.5)

Chapter 8

◆ Could it happen again? What was the worst that could have happened?

◆ Were there procedures in place? Were they followed? Were they good enough?

◆ Were those involved competent? What training and instruction had they been given?

◆ Could it have been picked up before it happened? If so, how?

Figure 8.5 Questions to be asked in an investigation.

A suitable accident report form is shown in Appendix 8.1.

☞ *Investigating Accidents and Incidents* HSG245, 2004, HSE Books, ISBN 9780717628278

8.6 Accident book

Under the Social Security (Claims and Payments) Regulations 1979, regulation 25, you, the employer must keep a record of accidents at premises where more than ten people are employed. Anyone injured at work is required to inform the employer and record information on the accident in an accident book including a statement on how the accident happened.

The employer must investigate the cause and enter this in the accident book if they discover anything which differs from the entry made by the employee. The purpose of this record is to ensure that information is available if a claim is made for compensation.

The HSE has produced an accident book BI 510 (ISBN 0-7176-2603-2) with notes on these Regulations and the Reporting of Injuries, Diseases and Dangerous Occurrences Regulations 1995 (RIDDOR), which are administered by the HSE (Fig. 8.6).

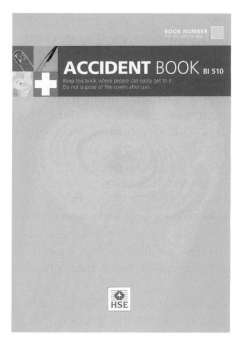

Figure 8.6 Accident book.

8.7 Reporting of accidents

8.7.1 General

RIDDOR 1995 requires the reporting of work-related accidents, diseases and dangerous occurrences. By giving this information you will enable the enforcing authorities to identify where and how risks arise and to investigate serious accidents. They can then advise you on preventive action to reduce injury, ill health and accidents.

RIDDOR requires you, if you are an employer, self employed or in control of premises, to report certain more serious accidents and incidents to the HSE or other enforcing authority and to keep a record. There are no

exemptions for small organisations. The reporting and recording requirements are as follows under Section 8.7.2.

8.7.2 What needs to be done?

In most cases, not very much – for most businesses a reportable accident,
dangerous occurrence or case of disease is a rare event.

The main points are that you must:

- ◆ Notify the enforcing authority immediately (e.g. by telephone) if
  anyone is killed or receives a major injury. This must be followed
  up within 10 days with a completed accident report form (F2508),
  available from the HSE.
- ◆ Notify the enforcing authority immediately if there is a dangerous
  occurrence. A dangerous occurrence is something that happens, that
  does not result in a reportable injury, but which clearly could have
  done. This must be followed up with an accident report form within
  10 days.
- ◆ Report within 10 days on form F2508, any injuries which keep an
  employee off work or unable to do their normal job for more than
  3 days.
- ◆ Complete a disease report form (F2508A) and send to the enforcing
  authority, if a doctor notifies the business that an employee is suffering from a reportable work-related disease.
- ◆ Keep a record of any reportable injury, disease or dangerous occurrence. This must include:
  - ○ date and method of reporting, for example by telephone or
    internet;
  - ○ date, time and place of event;
  - ○ personal details of those involved; and
  - ○ brief description of the nature of the event or disease.

To obtain more detailed information on what exactly constitutes a major
injury, dangerous occurrence or disease, you should look in the accident
book.

A convenient way of keeping this record is to use the accident book
produced by the HSE, which has details of what should be reported.
It also satisfies the requirement of The Social Security legislation to keep
a record of all injury accidents.

Instead of notifying the local office employers can notify the Incident Contact Centre. Telephone 0845 300 9923, website: www.riddor.gov.uk, fax 0845 300 9924 or e-mail: riddor@natbrit.com

☞ *RIDDOR Explained HSE 31Rev*

☞ *The Incident Contact Centre Misc. 310*

8.8 First aid

8.8.1 General

As an Employer you must ensure that adequate first-aid facilities are provided in the workplace and you need to be able to deal with situations which may arise requiring first aid. It is important that people can be treated immediately and that an ambulance can be called in serious cases.

The requirements for providing first aid at work depend on how risky a business is. In order to determine this, first carry out a risk assessment. Low risk businesses such as offices would only need to provide qualified first-aiders if there are over 50 employees. For other higher risk businesses such as a garage it may be necessary to provide first-aiders when there are less than 50 employees. There is no fixed guidance on the exact ratio. You need to think about how much risk there is to your employees. Table 8.1 is taken from the HSE's guide and gives their suggested numbers. If you are still unsure about what first-aid facilities are required, contact the local enforcing authority.

As a minimum an organisation must have:

- A suitably stocked first-aid box (see Section 8.8.3).
- An 'appointed person' (see Section 8.8.2) who can take charge of first-aid facilities and in an emergency. Emergency first-aid half-day training courses are available. An appointed person must be available whenever people are at work.
- Notices displaying the appropriate first-aid symbol, telling people where to find the first-aid box and who the appointed person is (Fig. 8.7).

Where there are special risks such as work with hazardous substances and dangerous tools and machinery:

- A specially trained first-aider and suitable equipment held in a first-aid room.

Chapter 8

Table 8.1 Suggested numbers of first-aid personnel

Category of risk	Numbers employed at any location	Suggested number of first aid personnel
Lower risk, for example, shops and offices, libraries	Fewer than 50	At least one appointed person
	50–100	At least one first-aider
	More than 100	One additional first-aider for every 100 employed
Medium risk, for example, light engineering and assembly work, food processing, warehousing	Fewer than 20	At least one appointed person
	20–100	At least one first-aider for every 50 employed (or part thereof)
	More than 100	One additional first-aider for every 100 employed
Higher risk, for example, most construction, slaughterhouses, chemical manufacture, extensive work with dangerous machinery or sharp instruments	Fewer than 5	At least one appointed person
	5–50	At least one first aider
	More than 50	One additional first-aider for every 50 employed

Source: HSE.

As your business grows, you must review the need for qualified first-aiders and the range of equipment required. First-aiders must have approved training (normally a 4 day course) and are only given a certificate valid for 3 years or less – after that they will need a refresher course and re-examination. Organisations who can provide training are registered with the Employment Medical Advisory Service (EMAS) and include organisations like the British Red Cross and St John's Ambulance.

Figure 8.7 First aid sign.

8.8.2 Appointed person

An appointed person is someone you choose to:

- take charge when someone is injured or falls ill, including calling an ambulance if required;
- look after the first-aid equipment, for example, restocking the first-aid box.

Appointed persons should not attempt to give first aid for which they have not been trained, though short emergency first-aid training courses are available. Remember that an appointed person should be available at all times people are at work on site – this may mean appointing more than one.

8.8.3 Contents of first-aid box

There is no standard list of items to put in a first-aid box. It depends on what you assess the needs are. However, as a guide, and where there is no special risk in the workplace, a minimum stock of first-aid items would be:

- a leaflet giving general guidance on first aid, for example, HSE leaflet *Basic advice on first aid at work* (see 'Where can I get further information?');
- Twenty individually wrapped sterile adhesive dressings (assorted sizes);
- two sterile eye pads;
- four individually wrapped triangular bandages (preferably sterile);
- six safety pins;
- six medium-sized (approximately 12 cm × 12 cm) individually wrapped sterile unmedicated wound dressings;
- two large (approximately 18 cm × 18 cm) sterile individually wrapped unmedicated wound dressings; and
- one pair of disposable gloves.

Chapter 8

You should not keep tablets or medicines in the first-aid box.

The above is a suggested contents list only; equivalent but different items will be considered acceptable.

☞ *First aid at work, your questions answered* INDG214

☞ *Basic advice on first aid at work* INDG347

8.9 Role and powers of enforcement officers

Inspectors from the Health and Safety Executive (HSE) (for factories, farms and building sites), or your local authority, usually Environmental Health Officers, (offices, shops, hotels, catering and leisure) cover the HSW Act. Officers from the local Fire and Rescue Authority enforce Fire Precautions under the Fire Safety Order (FSO).

Inspectors visit workplaces to check that people are sticking to the rules. They investigate some accidents and complaints but mainly they are there to help you understand what you need to do. They enforce only when something is seriously wrong.

Inspectors have significant powers to enforce the law. They have powers to enter your premises at any reasonable time (or any time in a dangerous situation) and to:

- ◆ examine and investigate;
- ◆ require premises to remain undisturbed;
- ◆ take measurements, photographs and recordings;
- ◆ cause an article or substance to be dismantled or subjected to tests;
- ◆ take possession of things for examination or legal proceedings;
- ◆ take samples;
- ◆ require people to give information and sign a statement;
- ◆ require information, facilities records or assistance;
- ◆ issue an IMPROVEMENT NOTICE to require safety improvements to be made;
- ◆ issue a PROHIBITION NOTICE to prohibit the use of a process, premises, machine, etc.
- ◆ initiate prosecutions against companies or individuals; and
- ◆ seize, destroy or render harmless any article or substance which is a source of imminent danger.

Under the Fire Safety Order, inspectors' powers are very similar. Under the Order they also have the use of an ALTERATIONS NOTICE to require people to notify the authority if alterations are made to the premises which could affect fire safety. In the FSO the Improvement Notice of the HSW Act becomes an ENFORCEMENT NOTICE.

8.10 Insurance claims

Accidents arising out of your company's activities resulting in injuries to people can lead to compensation claims. The second objective of an investigation should be to collect and record relevant information for the purposes of dealing with any claim. It must be remembered that prevention is the best way to reduce claims and must be the first objective in the investigation. An overzealous approach to gathering information concentrating on the compensation aspect can, in fact, prompt a claim from the injured party where there was no particular intention to take this route before the investigation. Nevertheless, relevant information should be collected. Sticking to the collection of facts is usually the best approach.

Usually the person claiming will go and seek the advice of a solicitor who must now follow fairly recent new procedures.

If you receive a letter of claim you need to act fairly quickly as the procedure involves:

1. 'Letter of claim' to be acknowledged within 21 days.
2. Ninety days from date of acknowledgement to either accept liability or deny. If liability is denied then full reasons must be given.
3. Agreement to be reached using a single expert.

The over riding message is that to comply with the protocols (arrangements) quick action is necessary. It is also vitally important that records are accurately kept and accessible.

By law (Employers' Liability (compulsory Insurance) Act 1969) you must have insurance for accident claims made against you for an injury or ill health from work activities. You should contact your insurance company immediately on receipt of a letter of complaint from an injured person and they will deal with the claim.

It is not expected that all accidents and incidents will be investigated in depth and a dossier with full information prepared. Judgement has to be applied as to which incidents might give rise to a claim and when a full record of information is required. All accident report forms should include the names of all witnesses as a minimum. Where the injury is likely to give rise to lost time, a photograph(s) of the situation should be taken. The kind of information that would be required is as follows:

◆ Accident book entry.

◆ First-aider report.

◆ Surgery record.

◆ Foreman/supervisor accident report.

◆ Safety representatives accident report.

◆ RIDDOR report to HSE/Local authority.

◆ Other communications between defendants and HSE.

◆ Minutes of Health and Safety Committee meeting(s) where accident/ matter considered.

◆ Report to DSS.

◆ Documents listed above relative to any previous accident/matter identified by the claimant and relied upon as proof of negligence.

◆ Earnings information where defendant is employer.

Documents produced to comply with requirements of the Management of Health and Safety at Work Regulations 1999:

◆ Pre-accident Risk Assessment.

◆ Post-accident Re-Assessment.

◆ Accident Investigation Report prepared in implementing the requirements; Health Surveillance Records in appropriate cases; Information provided to employees.

◆ Documents relating to the employees' health and safety training.

Appendix 8.1
Manager's incident/accident report

INJURED PERSON: ....Date of Accident: / /200   Time...........am/pm

POSITION: ....Place of Accident: ....

DEPARTMENT: .... Details of Injury: ....

Investigation carried out by: ....

Position: ....   Estimated Absence: ....

Brief details of Accident (A detailed report together with diagrams, photographs and any witness statements should be attached where necessary. Please complete all details requested overleaf.)

Immediate Causes	Underlying or Root Causes

Conclusions (How can we prevent this kind of incident/accident occurring again?)

Action to be taken: ....Completion Date: / /200

Recommendations for payment to be made to the injured employee for any sickness absence due to the above accident at work: **(Tick box as appropriate)**

❑ Basic Company Sick Pay
❑ Normal Daily Earnings (max.10 hours) for part week absence
❑ Nil Payment

IMPORTANT

1. Payment of Company Sick Pay is discretionary and applies to Company employees only. Payment recommendations above Basic Company Sick Pay Must be discussed and agreed with the Personnel Manager.
2. Please ensure that an accident investigation and report is completed and forwarded to Personnel within *48 hours of the accident occurring.*

Remember that accidents involving major injuries or dangerous occurrences have to be notified immediately by telephone to the local authority.

Signature of Manager making Report: .... Copies: Personnel Manager

Health & Safety Manager

Date: / /200   Payroll Controller

INJURED PERSON: Surname ....Forenames .... Male/Female

Home address .... Age............

Employee ☐ Agency Temp ☐ Contractor ☐ Visitor ☐ Youth Trainee ☐ (Tick one box)

Kind of Accident Indicate what kind of accident led to the injury or condition (tick one box)

Contact with moving machinery or material being machined ☐ 1	Injured whilst handling lifting or ☐ 5	Drowning or asphyxiation ☐ 9	Contact with electricity or an electrical discharge ☐ 13
Struck by moving including flying, or falling object ☐ 2	Slip, trip or fall on same level ☐ 6	Exposure to or contact with harmful ☐ 10	Injured by an animal ☐ 14
Struck by moving vehicle ☐ 3	Fall from height indicate approx. Distance of fall.........mtrs ☐ 7	Exposure to fire ☐ 11	Violence ☐ 15
Struck against something fixed or stationary ☐ 4	Trapped by something collapsing or overturning ☐ 8	Exposure to an explosion ☐ 12	Other kind of accident ☐ 16

Detail any machinery, chemicals, tools etc. involved

Accident first reported to: Name ....

Position & Dept..

First Aid/medical attention by: First Aider Name .... Dept ....
                            Doctor Name ....
                            Medical centre .... Hospital.....................

WITNESSES

Name	Position & Dept	Statement obtained (yes/no) Attach all statements taken
....	....	.... yes/no
....	....	.... yes/no
....	....	.... yes/no
....	....	.... yes/no

For Office use only

If relevant: Date reported to Enforcing Authority   a) by telephone ...../...../200
                                                b) on form F2508 ...../...../200

         Date reported to Company Insurers ...../...../200

Were the Recommendations Effective?   Yes/No
If No say what further action should be taken ....
....
....
....

Chapter 8

Appendix 8.2

Example of a shop emergency procedure

IMPORTANT TELEPHONE NUMBERS

- Ambulance...
- Fire...
- Police..
- Local Hospital/Doctor..
- Environmental Health Officer..
    - Name..
    - Address..
    - Telephone Number..

Emergency arrangements

In the event of an emergency created by a fire or a bomb scare or any other such incidents, an evaluation may become necessary.

It is essential that all team members are familiar with evacuation procedures. Team members have a responsibility to themselves and to others in becoming familiar with fire fighting equipment, fire doors, fire corridors, fire alarm operating points and the location of the First Aid Kit.

There should be an agreed meeting point, under cover if possible, where a roll call will be effected in the event of an emergency evacuation, all team members must be aware of this point. Evacuation procedures should be practised from time to time as part of a shop training session.

The Shop Manager, or in their absence the most senior member of staff present should assume responsibility for the successful evacuation of the shop premises and should, if necessary, appoint duties to each team member to assist the speed of the evacuation.

The Shop Manager, or in their absence the most senior member of staff present, is responsible for:

- Locking the shop when evacuating the premises.
- Checking that the till is locked and keeping the till keys in their possession (do not attempt to remove the contents of the till).
- Liaising with Centre Management (where applicable).

- Telephoning the emergency services (where applicable).
- Keeping Head Office informed of the situation.

Evacuation arrangements

These should be supervised, as listed above.

Escort members of the public and team members to the meeting point via the nearest exit or if necessary the fire exit. Try to minimise delay by preventing people from collecting personal possessions. Check that all the doors are closed and that no one is left inside the premises.

Bomb scares

In the event of a Bomb Alert being advised by the Police. or the Centre Management, evacuate the premises as for any other emergency, if advised to do so.

If a request is made to search the premises, do so thoroughly, being most vigilant in corners, displays and other unusual locations.

In the event of finding a package **DO NOT PANIC. DO NOT MOVE THE ITEM.** Leave the area and contact the Police and the Centre Management and let them deal with the package.

FIRE PRECAUTIONS AND PROCEDURES

On discovering a fire immediately operate the nearest fire alarm call point, and then tackle the fire if *possible* with the equipment provided but do not take any personal risk in doing so.

On hearing a fire alarm:

- Dial 999 to call the fire brigade.
- Leave the building by the nearest route.
- Close all doors behind you.
- Report to the person in charge of the assembly point.
- Do not take risks.
- Do not stop to collect personal belongings.
- Do not re-enter the building.

Appendix 8.3

Typical Fire Action Notice

Fire action
If you discover a fire

Operate nearest fire alarm point.

Call the Fire Brigade by telephoning ▢ 999

Leave the building by the nearest exit.

Report to your assembly point at ▭

Do not stop to collect personal belongings

Do not use lift

Sources of information and guidance

9

Summary

In this chapter:

- **A list of websites for principal sources of information and guidance**

- **A list of HSE and other short publications which are suitable for further reading**

- **Books by the same author**

- **Main acronyms used in health and safety**

9.1 Principal health, safety and fire websites

The authors and publishers are not responsible for the contents or reliability of the linked websites and do not necessarily endorse the views expressed within them. Listing should not be taken as endorsement of any kind. We cannot guarantee that these links will work all of the time and we have no control over the availability of the linked pages. The information is correct as of 31st January 2008.

◆ **Asbestos Removal Contractors Association (ARCA)** – www.arca.org.uk

ARCA is an association of specialist contractors committed to the safe removal of asbestos and other hazardous materials. It is a trade association representing all those professionally involved with the removal and abatement of asbestos. The association runs a range of training courses and seminars as well as publishing guidance notes. The website provides an on-line directory of all members categorized.

◆ **Association of Noise Consultants (ANC)** – www.association-of-noise-consultants.co.uk

The range of expertise available from members of the association includes hearing conservation and industrial noise control programs. Some members additionally provide educational courses and/or test laboratory facilities. The ANC produces a range of information sheets including one on how to comply with the Noise at Work Regulations.

◆ **Association of Specialist Fire Protection (ASFP)** – www.asfp.org.uk

The Association represents over 50 of the UK's major Manufacturers and Contractors as well as regulatory and certification bodies, involved in specialist passive fire protection. In order to promote quality, consistency and good working practices ASFP has a code of practice for the design, supply and installation of passive fire protection systems. ASFP and its members are fully committed to sustaining through performance their recognised position in the construction industry.

◆ **British Approvals for Fire Equipment (BAFE)** – www.bafe.org.uk

BAFE was established in 1984 to raise and maintain high standards of quality in active fire protection products and services. Firms registered with BAFE are all ISO 9000 certificated.

◆ **British Fire Protection Systems Association (BFPSA)** –
www.bfpsa.org.uk

The British Fire Protection Systems Association is the Trade
Association of manufacturers and installers of fire alarm and fixed
extinguishing systems. It is recognised as the *co-ordinating body for
the UK fire systems industry* and is the leading trade association in *its
field in Europe.*

◆ **British Occupational Hygiene Society (BOHS)** – www.bohs.org

BOHS is a Learned Society and organises conferences and publications for those with an interest in Occupational Hygiene.

◆ **Bully OnLine** – www.successunlimited.co.uk

Bully OnLine provides insight and guidance on identifying and
dealing with bullying at work and related subjects.

◆ **Chartered Institute of Wastes Management (CIWM)** –
www.ciwm.co.uk

The IWM is the professional body which represents over 4,000 waste
management professionals – predominantly in the UK but also over-
seas. The IWM sets the professional standards for individuals working in the waste management industry and has various grades of
membership.

◆ **Chemical Hazards Communication Society (CHCS)** –
www.chcs.org.uk

CHCS is the active professional body for individual membership by
all those involved in the areas suggested by its name. It holds seminars and training courses in relevant areas and publishes a series
of Safety Data Sheets ('SDSs') and A User Guide to assist those who
receive SDSs.

◆ **Confederation of British Industry (CBI)** –
www.cbi.org.uk

The CBI's objective is to help create and sustain the conditions in
which the UK can compete and prosper. Through its network of
offices around the UK and in Brussels, it represents its members'
views on all cross-sectoral issues to the government and other
national and international policy-makers. It supplies advice, information and research services to members on key public policy issues
affecting business and provides a platform for the exchange and
encouragement of best practice.

● **Considerate Constructors Scheme** – www.ccscheme.org.uk

The British public has traditionally taken a dim view of construction. The industry is often seen as uncaring and insensitive. Construction sites are viewed as noisy, dirty and dangerous places. People often feel reluctant to complain, fearing at best their complaint will be ignored, or at worst that it prompts an abusive response. The Considerate Constructors Scheme provides a real opportunity for the industry to improve its public image. It has the full support of every sector of construction and is backed by the Government, who recognise the importance of the demonstrating consideration for both the public and the environment.

● **Department for Communities and Local Government (DCLG)** – www.communities.gov.uk/fire/firesafety/firesafetylaw/ and also can be ordered at www.firesafetyguides.co.uk

DCLG has responsibility for Fire and has published a series of guides on the Regulatory Reform Fire Safety Order 2005 which can be down loaded free from their website.

● **Department for Rural Affairs (DEFRA)** – www.defra.gov.uk

Defra (the Department for Environment, Food and Rural Affairs) works for the essentials of life-food, air, land, water, people, animals and plants. Their remit is the pursuit of sustainable development-weaving together economic, social and environmental concerns.

● **Department for Transport (DFT)** – www.dft.gov.uk

This is a recently new Department for Transport to focus solely on transport issues. The department has responsibilities for aviation, integrated transport, local transport, mobility and inlcusion, roads, railways and shipping.

● **Department for Business Enterprise and Regulatory Reform (BERR)** – www.berr.gov.uk

The Department brings together functions from the former Department of Trade and Industry, including responsibilities for productivity, business relations, energy, competition and consumers, with the *Better Regulation Executive (BRE)*, previously part of the Cabinet Office (see also the *Better Regulation website* to submit ideas for reducing or simplifying regulation for the business, public and third sectors).

Chapter 9

◆ Environment Agency – www.environment-agency.gov.uk

As 'Guardians of the Environment', the Environment Agency has legal duties to protect and improve the environment throughout England and Wales and so contribute towards 'sustainable development'-meeting the needs of today without harming future generations. The Agency is officially a 'non-departmental public body' which means that they work for the public and have specific duties and powers of their own. Their main sponsor in the Government is the Department of Environment, Transport and the Regions (DETR). They also have links to the Ministry of Agriculture, Fisheries and Food and the Welsh Office.

◆ European Agency for Safety and Health at Work – www.uk.osha.europa.eu

The European Agency for Safety and Health at Work was set up by the European Union to promote the exchange of technical, scientific and economic information between member states. In the UK, Health and Safety Executive (HSE) acts as the main point of contact for the European Agency to enable health and safety at work information gathered throughout the EU to be available in the UK.

◆ FIRA International Ltd – www.fira.co.uk

FIRA has driven the need for higher standards through testing, research and innovation for the furniture and allied industries. A non-government funded organisation, FIRA is supported by all sections of the industry through the Furniture Industry Research Association, ensuring ongoing research programmes which bring benefits to all. With unrivalled support from across the whole industry, FIRA also has the influence and capability to help shape legislation and regulations.

◆ Fire Protection Association (FPA) – www.thefpa.co.uk

The Fire Protection Association is the UK's national fire safety organisation. The association works to identify and draw attention to the dangers of fire and the fire prevention measures.

◆ Forestry Commission – www.forestry.gov.uk

The *Forestry Commission* is the Government department responsible for forestry throughout Great Britain. Its mission is to protect and expand Britain's forests and woodlands and increase their value to society and the environment.

Chapter 9

◆ Hazards Forum – www.hazardsforum.co.uk

The Hazards Forum was established in 1989 by the four major engineering institutions, the Institutions of Civil, Electrical, Mechanical and Chemical Engineers. The Forum was set up because of concern about several major disasters, both technological and natural. The Forum believes that there is widespread public misunderstanding of the nature of hazards and the risks they pose. The public reaction to the risks from genetically modified foods and those from smoking is one example of differing attitudes. These attitudes may result from media attention; suspicion of assurances of safety bodies with a vested interest; or just ignorance. Whatever the reason the Forum believes there is a need to fill an educated but unbiased role and this the Forum aspires to do.

◆ Health and Safety Executive (HSE) – www.hse.gov.uk

The HSE's main website containing general information on their areas of work and the latest news.

◆ Health and Safety Laboratory (HSL) – www.hsl.gov.uk

HSL is Britain's leading industrial health and safety facility, and offers a unique portfolio of skills and expertise. Operating as an agency of the Health and Safety Executive (HSE), HSL plays a pivotal role in supporting HSE's mission to ensure that risks to people's health and safety from work activities are properly controlled. This involves HSL in two main areas of activity: operational support through incident investigations and studies of workplace situations; and longer term work on analysis and resolution of occupational health and safety problems.

◆ HSE Bookfinder – www.hsebooks.co.uk

An online catalogue of HSE publications available from HSE books.

◆ HSE Public Register of Convictions – www.hse-databases.co.uk/prosecutions

This site gives details of all prosecution cases taken by HSE, since 1 April 1999, which resulted in a conviction.

◆ Incident Contact Centre – www.hse.gov.uk/riddor

RIDDOR stands for the Reporting of Injuries, Diseases and Dangerous Occurrences Regulations 1995. This site provides information about RIDDOR and allows you to report accidents, diseases and dangerous occurrences. This new arrangement is available for all incidents that occur on or after 1 April 2001.

- ◆ **Institution of Occupational Safety and Health (IOSH)** – www.iosh.org.uk

  IOSH is the chartered organization representing occupational safety and health practitioners predominately in the UK but elsewhere also. IOSH have a valuable website with technical services for members; training courses and many local branch meetings for practitioners.

- ◆ **Local Government Association (LGA)** – www.lga.gov.uk

  The LGA provides the national voice for local communities in England and Wales; its members represent over 50 million people, employ more than 2 million staff and spend over £65 billion on local services. We work with and for our member authorities to realise a shared vision of local government that enables local people to shape a distinctive and better future for their locality and its communities.

- ◆ **National Examination Board in Occupational Safety and Health (NEBOSH)** – www.nebosh.org.uk

  NEBOSH is the main QCA accredited awarding body for Level 6 and Level 3 health and safety qualifications. Their awards are run by over 400 training organizations.

- ◆ **Passive Fire Protection Federation (PFPF)** – www.associationhouse.org.uk

  Federation of trade associations, test houses, certification and government bodies involved in passive fire protection.

- ◆ **Royal Society for the Prevention of Accidents (RoSPA)** – http://www.rospa.org.uk

  RoSPA's mission is to save lives and reduce injuries. The Royal Society for the Prevention of Accidents is a registered charity established over 80 years ago and aims to campaign for change, influence opinion, contribute to debate, educate and inform – for the good of all. By providing information, advice, resources and training, RoSPA is actively involved in the promotion of safety and the prevention of accidents in all areas of life – at work, in the home, and on the roads, in schools, at leisure and on (or near) water.

- ◆ **Safety Assessment Federation (SAFed)** – www.safed.co.uk

  SAFed is the leading trade body for the independent engineering safety inspection and assessment industry. Between them, SAFed member companies employ over 2200 inspection staff with specialist expertise covering an extensive range of mechanical and electrical plant, equipment and systems.

Chapter 9

◆ Safety Health and Environment Intra Industry Benchmarking Association (SHEIIBA) – www.sheiiba.com

SHEIIBA offers web based benchmarking tools designed for intra company knowledge sharing and performance comparisons.

◆ Scottish Environment Protection Agency (SEPA) – www.sepa.org.uk

SEPA's main aim is to provide an efficient and integrated environmental protection system for Scotland that will both improve the environment and contribute to the Scottish Ministers' goal of sustainable development.

◆ Trade Unions Congress (TUC) – www.tuc.org.uk

The TUC is the voice of Britain at work with 76 member unions representing 6.8 million working people from all walks of life. The TUC website includes press releases, reports, rights at work information and has an area devoted to health and safety issues.

◆ Trading Standards Institute – www.tradingstandards.gov.uk

The Trading Standards Institute exists to enhance the professionalism of its members in the Trading Standards Service in support of informing consumers, encouraging honest businesses and targeting rogue traders.

◆ Working Well Together – www.wwt.uk.com

The Working Well Together site is the online centre of excellence for information, advice and guidance for everybody involved in the construction industry who wants to improve standards of health and safety.

9.2 HSE books publications referenced in the text of various chapters and other booklets that are worth consulting

Most of these and more are available free or as a download from the HSE website: www.hse.gov.uk . There are many more thicker and more detailed guidance provided by HSE books. If you wish to study more information go to HSE books website: www.hse.co.uk or get their latest catalogue:

Chapter 9

Title	Stock code	Publisher, Publication year	ISBN-10/ ISBN-13
A guide for new and expectant mothers who work	INDG373	2003	ISBN 0-7176-2614-8
Aching arms (or RSI) in small businesses	INDG171(rev1)	2005	ISBN 0-7176-2600-8
Advice to managers on 'asbestos essentials' – Introduction to task sheets		HSE: www.hse. gov.uk/asbestos/ essentials/index.htm, 2007	
An introduction to health and safety	INDG259(rev1)	2003	
Asbestos essentials task manual: Task guidance sheets for the building maintenance and allied trades	HSG210	2001	ISBN 0-7176-1887-0
Back in work	INDG333	web only, 2000	
Basic advice on first aid at work	INDG347(rev1)	2006	ISBN 0-7176-619382261-84
Buying new machinery	INDG271	1998	ISBN 0-7176-1559-6
Checkouts and musculoskeletal disorders	INDG269	2002	ISBN 0-7176-1615-0539-1
Consulting Employees on Health and Safety	INDG232	1996	ISBN 0-7176-2170-7

(Continued)

Chapter 9

Title	Stock code	Publisher, Publication year	ISBN-10/ ISBN-13
Control back pain risks from whole-body vibration	INDG242(rev1)	2005	ISBN 0-7176-6119-9
COSHH a brief guide to the regulations	INDG136 (rev32)	2005	ISBN 0-7176-2982-1
Directors' responsibility for health and safety	INDG343	2002	ISBN 0-7176-2080-8
Don't mix it – a guide for employers on alcohol	INDG240	www.hse.gov.uk/ pubns/indg240.pdf, 2007	
Drive away bad backs	INDG404	2005	ISBN 0-7176-6120-2
Driving at work	INDG382	2003	ISBN 0-7176-2740-3
Drug misuse at work	INDG91	1998	0-7176-2402-1
Electrical safety and you	INDG231	2005	ISBN 0-7176-1207-4
Employers' Liability (Compulsory Insurance) Act 1969 – A guide for employers	HSE40(rev)	HSE website, 2002	
Essentials of health and safety at work		2006	ISBN 0-7176-6179-2
Fire – A short guide to making your premises safe from fire		www.firesafetyguides. cocommunities. gov.uk and www. communities.gov. uk/fire/firesafety/ firesafetylaw/	

(Continued)

Title	Stock code	Publisher, Publication year	ISBN-10/ ISBN-13
Fire and explosion	INDG370	2004	ISBN 0-7176-2589-3
Fire safety risk assessment – offices and shops	2006	Department for Communities and Local Government	ISBN 9780-1-85112-815-0
First-aid at work, your questions answered	INDG214	2006	ISBN 0-7176-1074-8
Five steps to risk assessment	INDG163(rev2)	2006	ISBN 0-7176-6189-X-1565-0
Getting to grips with manual handling	INDG143(rev2)	2004	ISBN 0-7176-2828-5
Hand-arm vibration advice for employers	INDG296(rev)	2006	ISBN 0-7176-6118-0
Health risks from hand-arm vibration	INDG175(rev2)	2005	ISBN 0-7176-6117-2
Health and Safety law – What you should know	2006	HSE website	ISBN 0-7176-1702-5
Health and safety training: What you need to know	INDG345	2001	ISBN 0-7176-2137-5
Home-working – Guidance for employers and employees on health and safety	INDG226	2006	ISBN 0-7176-1204-X
Idiot's guide to CHIP3	INDG350	2003	ISBN 0-7176-2333-5
Incident at work	MISC769	HSE website, 2007	

(Continued)

Chapter 9

Title	Stock code	Publisher, Publication year	ISBN-10/ ISBN-13
Legionnaires' disease – Essential information for providers of residential accommodation	INDG376	2003	C1000
Legionnaires' disease, A guide for employers	IACL27(rev2)	2004	ISBN 0-7176-1773-4
Maintaining portable electrical equipment in offices and other low risk environments	INDG236	2004	ISBN 0-7176-12723-42
Managing asbestos in premises	INDG223(rev3)	2004	ISBN 0-7176-2092-12564-8
Managing health and safety – Five steps to success	INDG275	1998	ISBN 0-7176-2170-7
Managing health and safety in construction: Approved Code of practice	L144	HSC, 2007	ISBN 978-0-7176-6223-4
Managing occupational road risk		Royal Society for the Prevention of Accidents available from Edgbaston Park, 353 Bristol Road, Birmingham B5 7ST Tel: 0121 248 2000	

(Continued)

Title	Stock code	Publisher, Publication year	ISBN-10/ ISBN-13
Managing sickness absence and return to work in small businesses	INDG399	2004	ISBN 0-7176-2914-7
Managing vehicle safety at the workplace	INDG199	2006	ISBN 0-7176-2821-3
Need help on health and safety?	INDG322	web only, 2000	ISBN 0-7176-1790-4
No second chances – A farm machinery safety step by step guide	INDG241	HSE website, 2005	
Noise at work a guide for employers	INDG362(rev1)	2005	ISBN 0-7176-6165-2
Officewise	INDG173	2006 (reprinted with updates)	ISBN 0-7176-1518-9
Passport schemes for health, safety and the environment	INDG381	2003	
Personal Protection at Work Regulations	INDG174(rev1)	1992, 2005	ISBN 0-7176-6141-50889-0
Pressure systems safety and you	INDG261	2001	ISBN 0-7176-1562-6
Preventing contact dermatitis at work	INDG233(rev1)	2007	ISBN 0-7176-6183-11246-5
Preventing slips and trips at work	INDG225(rev1)	2007	ISBN 0-7176-2760-8
Protect your hearing or loose it!	INDG363(rev1)	2007	ISBN 0-7176-6166-40

(Continued)

Chapter 9

Title	Stock code	Publisher, Publication year	ISBN-10/ ISBN-13
Rash decisions		HSE video on work related dermatitis – Its causes, effects and prevention	
Read the label	INDG352	2004	ISBN 0-7176-2367-X186,
Reduce risks cut costs	INDG355	2002	ISBN 0-7176-2337-8
RIDDOR explained	HSE31(rev)	1999	ISBN 0-7176-2441-2
Safe use of ladders and stepladders: An employers guide	INDG402	2005	ISBN 0-7176-6105-9
Safe work in confined spaces	INDG258	1999	ISBN 0-72176-1442-5
Safe working with flammable substances	INDG227	2005	ISBN 0-7176-1154-X
Signpost to The Health and Safety (Safety Signs and Signals) Regulations 1996	INDG184L	1996	ISBN 0-7176-1139-6
Simple guide to LOLER	INDG290	1999	ISBN 0-7176-2430-7
Simple guide to PUWER	INDG291	1999	ISBN 0-7176-2429-3
Smokefree guide for England		http://www. smokefreeeng land.co.uk	
Successful health and safety management	HSG65	1997	ISBN 0-7176-1276-7

(*Continued*)

Title	Stock code	Publisher, Publication year	ISBN-10/ ISBN-13
Sun protection	INDG337	2001	ISBN 0-7176-1982-6
Tackling work related stress, The Management Standards Approach	INDG406	2005	ISBN 0-7176-6140-7
The High 5 for small construction sites	INDG384	2003	
The Highway Code		Direct Government website, can be viewed on: www.highwaycode. gov.uk http://www. direct.gov.uk/en/ TravelAndTransport/ Highwaycode/index. htm, 2007	
The HSE and You	HSE37		
The incident contact centre	Misc310		
The Work at Height Regulations 2005 (as amended) a brief guide	INDG401(rev)	2007	ISBN 0-7176-6231-9
Use of contractors a joint responsibility	INDG368	2002	ISBN 0-7176-2566-4
Using work equipment safely	INDG229	2002	ISBN 0-7176-2389-0
Vibration – Control the risks from hand-arm vibration	INDG175	2005	ISBN 0-7176-6117-2
Violence at work	INDG69(rev)	1999	ISBN 0-7176-1271-6

Chapter 9

(Continued)

Title	Stock code	Publisher, Publication year	ISBN-10/ ISBN-13
Want construction work done safely: A guide for clients on the CDM2007 regulations	INDG411		ISBN 9780-7176-6246-3
Welfare at work	INDG293(rev1)	2007	ISBN 9780-7176-6264-7
Why do I need a safety data sheet?	INDG353	2002	ISBN 0-7176-2367-X
Work related stress	INDG281	2001	ISBN 0-7176-2112-X
Working alone in safety	INDG73(rev)	2005	ISBN 0-7176-1507-3
Working safely with solvents	INDG273	HSE website, 2003	ISBN 0-7176-2246-0
Working together to relieve stress at work: A guide for employees		www.hse.gov.uk/ pubns/stress.htm, 2004	
Working together	INDG268	web only, 2002	
Working with VDUs	INDG36(rev3)	2006	ISBN 0-7176-6222-5
Workplace exposure limits	EH40	2005	ISBN 0-7176-2977-5
Workplace transport safety an overview	INDG199(rev1)	2006	ISBN 0-7176-2821-3
Workplace, health, safety and welfare	INDG244(rev2)	2006	ISBN 0-7176-6192-X
Your health, your safety; A guide for workers	HSE27(rev)	HSE website, 2004	

Chapter 9

9.3 Books by Phil Hughes MBE and Ed Ferrett

Introduction to Health and Safety at Work, 3rd edition	Elsevier, 2007	ISBN 9780750685030
Introduction to Health and Safety in Construction, 2nd edition	Elsevier, 2006	ISBN 9780750681117

9.4 A few acronyms used in health and safety

ACOP	Approved Code of Practice
ACM	Asbestos containing material
CAR	Control of Asbestos Regulations
CBI	The Confederation of British Industry
CECA	The Civil Engineering Contractors Association
CDM	The Construction (Design and Management) Regulations
CHIP	Chemicals (Hazard Information and Packaging) Regulations
CIRA	Construction Industry Research and Information Association
CONIAC	Construction Industry Advisory Committee
COSHH	Control of Substances Hazardous to Health Regulations
dB(A)	Decibel (A-weighted)
dB(C)	Decibel (C-weighted)
DSE	Display Screen Equipment Regulations
DSEAR	Dangerous Substances and Explosive Atmospheres Regulation

EAV	Exposure action value
EC	European Community
ELV	Exposure limit value
EMAS	Employment Medical Advisory Service
EPA	Environmental Protection Act 1990
EU	European Union
HAV	Hand-arm vibration
HSC	Health and Safety Commission
HSE	Health and Safety Executive
HSW Act	Health and Safety at Work etc. Act 1974
IOSH	Institution of Occupational Safety and Health
LEAL	Lower exposure action level
LOLER	Lifting Operations and Lifting Equipment Regulations 1998
MCG	The Major Contractors Group
MHOR	Manual Handling Operations Regulations
MHSW	Management of Health and Safety at Work Regulations
MORR	Management of Occupational Road Risk)
MoT	Ministry of Transport (still used for vehicle tests)
NAWR	Control of Noise at Work Regulations
NEBOSH	National Examination Board in Occupational Safety and Health
OHSAS	Occupational Health and Safety Assessment Series
OSH	Occupational Safety and Health

POOSH	Professional Organisations in Occupational Safety and Health)
PPE	Personal protective equipment
PUWER	The Provision and Use of Work Equipment Regulations
RIDDOR	The Reporting of Injuries, Diseases and Dangerous Occurrences Regulations
ROES	Representative(s) of Employee Safety
RoSPA	Royal Society for the Prevention of Accidents
RRFSO	Regulatory Reform Fire Safety Order
RSI	Repetitive Strain Injury
SGUK	Safety Groups UK
SPL	Sound Pressure Level
STEL	Short-term exposure limit
TUC	Trades Union Congress
UEAL	Upper exposure action level
UK	United Kingdom
VAWR	Vibration at Work Regulations
WAHR	Work at Height Regulations
WBV	Whole body vibration
WEL	Workplace Exposure Limit
WRULD	Work-related upper limb disorder

Chapter 9

Index